Dialogues on C

Written both for general readers and college students, *Dialogues on Climate Justice* provides an engaging philosophical introduction to climate justice, and should be of interest to anyone wanting to think seriously about the climate crisis.

The story follows the life and conversations of Hope, a fictional protagonist whose life is shaped by a terrifyingly real problem: climate change. From the election of Donald Trump in 2016 until the 2060s, the book documents Hope's discussions with a diverse cast of characters. As she ages, her conversations move from establishing the nature of the problem, to engaging with climate skepticism, to exploring her own climate responsibilities, through managing contentious international negotiations, to considering big technological fixes, and finally, as an older woman, to reflecting with her granddaughter on what one generation owes another. Following a philosophical tradition established by Plato more than two thousand years ago, these dialogues are not only philosophically substantive and carefully argued, but also distinctly *human*. The differing perspectives on display mirror those involved in real-world climate dialogues going on today.

Stephen M. Gardiner is Professor of Philosophy and Ben Rabinowitz Professor of the Human Dimensions of the Environment at the University of Washington, Seattle. He is the author of *A Perfect Moral Storm: The Ethical Tragedy of Climate Change* (2011), and co-author of *Debating Climate Ethics* (2016). His edited books include *The Ethics of "Geoengineering" the Global Climate* (2020), *The Oxford Handbook of Environmental Ethics* (2016), and *The Oxford Handbook of Intergenerational Ethics* (2023).

Arthur R. Obst is a PhD Candidate at the University of Washington, Seattle, who has lived his whole life in the shadow of climate change, a force that has brought not only environmental crisis but a conceptual crisis for environmentalists. He is dedicating his scholarship to addressing both.

Philosophical Dialogues on Contemporary Problems

Philosophical Dialogues on Contemporary Problems uses a well-known form – at least as old as Socrates and his interlocutors – to deepen understanding of a range of today's widely deliberated issues. Each volume includes an open dialogue between two or more fictional characters as they discuss and debate the empirical data and philosophical ideas underlying a problem in contemporary society. Students and other readers gain valuable, multiple perspectives on the problem at hand.

Each volume includes a foreword by a well-known philosopher, topic markers in the page margins, and an annotated bibliography.

Titles in series:

Dialogues on Ethical Vegetarianism
Michael Huemer

Dialogues on the Ethics of Abortion
Bertha Alvarez Manninen

Dialogues on Climate Justice
Stephen M. Gardiner and Arthur R. Obst

Forthcoming:

Dialogues on Free Will and Responsibility
Thomas A. Nadelhoffer

Dialogues on Gun Control
David DeGrazia

For more information about this series, please visit: www.routledge.com/Philosophical-Dialogues-on-Contemporary-Problems/book-series/PDCP

Dialogues on Climate Justice

Stephen M. Gardiner
and Arthur R. Obst

Routledge
Taylor & Francis Group

NEW YORK AND LONDON

Cover image: Lynn Gardiner

First published 2023
by Routledge
605 Third Avenue, New York, NY 10158

and by Routledge
4 Park Square, Milton Park, Abingdon, Oxon, OX14 4RN

*Routledge is an imprint of the Taylor & Francis Group, an informa
business*

Library of Congress Cataloging-in-Publication Data
A catalog record for this title has been requested

ISBN: 978-0-367-64196-2 (hbk)
ISBN: 978-0-367-64195-5 (pbk)
ISBN: 978-1-003-12340-8 (ebk)

DOI: 10.4324/9781003123408

Typeset in Sabon LT Std
by KnowledgeWorks Global Ltd.

Contents

Preface

Every book is written in a context. Ours took shape late in the summer of 2020 in Seattle. Living in the Pacific Northwest, we knew to cherish the glorious, bright days before the grey of winter returns. But that year gloom descended early. Wildfires brought thick smoke, wrapping the city in shades of grey, white, and sometimes orange for days at a time.

The U.S. Air Quality Index (AQI) runs from 0 (good) to over 300 (hazardous), with several gradients of worrying in between. That September, many areas in California, Oregon, Washington, and British Columbia were officially pronounced either "unhealthy" (151–200), "very unhealthy" (201–300), or "hazardous" (over 300), with authorities urging residents to stay indoors with windows shut.[1] Near Portland, the smog became so thick that some neighborhoods had days well over 400 AQI, and occasional periods *above* 500,[2] raising questions as to whether new, even more serious designations are needed *beyond* "hazardous".

The effects on the ground were neither statistical nor abstract, but *human*. In California, five of the six largest fires since 1932 burned simultaneously, overwhelming firefighters, forcing residents to flee, destroying tens of thousands of structures, laying waste to 4 million acres, and sending a great plume of smoke across the Great Plains.[3] Some lost their homes. Some died. Many,

1 Farmer & Chiquiza, 2020.
2 Samayoa, 2020.
3 Center for Disaster Philanthropy, 2020.

already confined by the ongoing Covid-19 pandemic, now couldn't even open a door or window for fear of letting in the smoke. The West, long a symbol of freedom, felt profoundly claustrophobic.[4]

On some days, our outlooks were as bleak as they have ever been. Writing in a tumultuous time, with serious climate action seeming far out of reach, we worried that the world was closing in on a future that Steve had tried to prevent and Arthur long feared his generation would have to endure. We could feel climate change in our bones. *How did it come to this? How did it go so wrong?*

By March of 2021, as we completed the first full draft of the text, the sun shone brightly in a baby blue sky as Seattle awakened with the first signs of spring. The haze was gone, at least for a time. A new President, Joe Biden, had recently taken office, elected in part on a platform that promised far more aggressive action on climate than ever seen before in this country. On his first day, the President signed an executive order enabling the United States to rejoin the Paris Climate Accords. He also repeated his commitment to achieving a net-zero economy by 2050, and to addressing environmental injustice, vowing to put it "at the center of all we do."[5]

In those moments, the prospects for robust climate action looked much brighter, and some in the climate movement were relishing what seemed like a big win, one that would put the country and the world on the right track. For many, the gloom we felt not half a year ago seemed to have blown away with the smoke and winter clouds. One could sense the hope in the air that later this century the most powerful question might become: How did we turn it around? After pushing so close to the edge, *how did it then go so right?*

Hope was in our hearts too. Still, we could not forget that such feelings are fickle friends. Real excitement had bubbled up before, with events such as the creation of the United Nations Framework Convention on Climate Change in 1992, the popularity of Al Gore's *An Inconvenient Truth* in 2001, President Obama's climate promises in 2008 and 2012, and the Paris Agreement in 2015.

4   Marris, 2020.
5   The White House, January 27, 2021.

Sadly, the reality is that all of these were false dawns, moments of apparent climate progress that might be built on, but which remain embedded within an overall trajectory of ongoing failure.

Was the new moment different? We desperately wanted it to be. Nevertheless, we remain convinced that the journey is long, and we are closer to the beginning than to the end. There will be many struggles ahead in the quest for a just climate future. Tribulations will come again, as surely as late season wildfires in California or Seattle's winter rains; but, so too will victories. To weather the turning tides of justice, we suggest, one must navigate with eyes on the horizon, stay focused on where you need to go, and do what you can to help the boat along. If no single leader turns out to be the climate savior, that is because there is no single climate savior. Instead, the focus should be on a whole world of people who, together, have the power to determine what kind of future they, their descendants, and the rest of nature will come to know.

We hope this book will help the reader to navigate a path forward, with the knowledge to do their part in making this future better, and the inspiration to bring us closer to our dreams than our nightmares. May the future look back, half a century or more from now, and tell a tale of what went right.

Foreword

I first became aware that humans might be changing the climate in 1980. I had just arrived at the University of Colorado, Boulder, and part of my job was to teach an environmental ethics course with a climate scientist who worked at the National Center for Atmospheric Research (NCAR). Mickey Glantz (the scientist) was primarily interested in extreme events and how they ramified through an unjust and tragic world. Through Mickey, I soon got to know other NCAR scientists who were more directly concerned with how future climates could be inferred from the output of supercomputers crunching wholly inadequate data sets.

I found these issues fascinating, but of marginal relevance to both the world and to my career. Like many people, I thought of the Earth as a self-regulating system: if carbon dioxide were forcing the climate in one direction, there would probably be some as yet undiscovered feedback that would bring it back into equilibrium. Even if this were not true, then surely the adults who run the world would do something to fix the problem before things got out of hand. Boy, was I wrong. As for my career, I was a serious philosopher, and thinking about "the weather" was kind of an illicit hobby. But four decades later, climate change has become central to what I do, more as a seeming inevitability than a matter of choice. The way that I see it, rather than abandoning philosophy in order to think about climate, I couldn't help but go beyond philosophy.

Climate change is now a huge area of academic study, and climate ethics and justice are part of this larger domain. This to a great extent is due to Steve Gardiner's 2011 landmark study, *A Perfect Moral Storm*. Despite the book's influence, some of its most important lessons have not been taken up. Many people who talk about climate justice and ethics seem to think that it is enough to know who the bad guys are and hold them accountable. They sometimes seem oblivious to the complexity and profundity of the issues. At the same time, some professional philosophers and political theorists are so concerned with their fields of academic expertise that they neglect the political and cultural realities in which arguments about climate ethics and justice occur. If we care about the world, it is not enough to know what some philosopher thinks is the best argument. What matters more is what we can be brought to believe and do together.

Dialogical thinking and writing is made for this challenge. Philosophy was born in dialogue, and what remains from Socrates (its founder in the West) is not the treatises he wrote (he didn't write any) or the articles he published in refereed journals (there weren't any), but the dramatized record of conversations he had with whoever was willing to talk to him. Professional philosophers today tend to think of their real work as writing books and articles (exactly what Socrates didn't do). Yet what most philosophers actually do is remarkably similar to what Socrates did: they talk to people. They try to clarify their thoughts, point out inconsistencies, show them the consequences of their beliefs, and generally help them through the mental labor of reaching conclusions that can inform how they should live (Socrates' analogy to the philosopher as midwife remains apt). Working through these questions can be an arduous process requiring a great deal of patience, and so at odds with much of what passes for conversation in the contemporary world. You can't really have much of a conversation on Twitter, and sound bites and social media posts do not often lead to livable conclusions, or even mutual respect, much less consensus.

If anything like climate justice is to be achieved, the conversation has to widen and its quality improve. While contemporary philosophers do not deal with ancient Athenian elites, the people they mostly talk to are the modern elites who study in colleges and universities, whose norms and values seem peculiar, eccentric, or even oppressive to some outside of these institutions. Progress on climate justice requires the participation of more than the wise and educated denizens of colleges and universities and the marginalized communities that have caught their attention. It also requires the participation of those who do not see themselves as "winning bigly" from the current order of things, but just as people trying to hang on to what they have while the ground is shaking beneath them.

Perhaps the most difficult obstacle to overcome in discussing climate change, even with those who are open-minded and good-hearted, concerns its collective nature and temporal and geographic scale. We are a solution-centric people. The problems that we are interested in are those that we can fix, here and now. Once people understand that addressing climate change is not just a matter of better batteries or international agreements with "greater ambition," there is a tendency to sink into apathy. One of the most impressive features of this book is the way that Gardiner and Obst address this obstacle, both substantively and dramaturgically.

There is so much to like in this book that it seems almost arbitrary to talk about one thing in particular, yet I can't help myself. The decision to focus on a single character (the aptly named Hope) who moves through time and life-stages brings out brilliantly some of the personal challenges in living with climate change. Hope enters the story as a beginning law student, and we follow her throughout her career, and into retirement. We meet her fellow students, father, colleagues in government, and finally her granddaughter Caitlyn. It appears that Hope has lived a good life. When we see her, she is treating people with respect, yet she is persistent in her work on climate change and in keeping justice at the center. The world seems better because of Hope and her work. Yet Hope does not shrink from admitting that she is part of an immoral generation, and that she herself could and should have done more. In her old age she finds herself living in a world that she had tried to prevent; still she hopes that her granddaughter will have a good life and be part of a generation that will overcome the obstacles that Hope's generation could not. Despite the failures, Hope has not sunk into despair or resignation. She takes pride in her granddaughter (and still enjoys her lemonade). The fact is, the most we can hope for on Caitlyn's behalf is that her life will have a broadly similar trajectory to her grandmother's, for the struggle to live meaningfully in a world of climate disruption will go on for centuries. Perhaps the most difficult truth about climate change to grasp and to live with is this: we will fail, yet it matters what we do. This book will not "save the world," for nothing can, and anyway it is us, not the world that most needs saving. What we can and should hope for is that this book makes us more like Hope.

Dale Jamieson
New York, NY

Acknowledgments

This text could not have been written without the support we have received. We are deeply honored to have a giant of climate ethics, Dale Jamieson, write the foreword. We are indebted to Jamie Draper, Clare Heyward, Marion Hourdequin, and Kian Mintz-Woo for their detailed comments on an earlier draft of the text, discussed at an online workshop hosted by the University of Washington in April 2021. We're also grateful to Thomas Ackerman, Roger Crisp, Matthew Gardiner, Sofia Huerter, Keith Hyams, Nic Jones, Catriona McKinnon, Colin Marshall, Jamie Mayerfield, Jonathan Milgrim, Guilio Reid-Thomas, Allen Thompson, Paul Tubig, and Kyle Powys Whyte for helpful suggestions. We thank students in our environmental ethics, justice, and contemporary moral problems courses, where we trialed some of the material. We also honor the countless climate scholars, leaders, and activists who have committed much of their lives to protecting the future. Although we could not possibly reference all of them by name, their influence runs through this book.

Steve thanks the Program on Ethics, the Department of Philosophy, the College of Arts and Sciences, and the College of the Environment at the University of Washington, Seattle, for their support. He is particularly grateful to the Rabinowitz family for their commitment to philosophy and to the environment. As one of the chapters was completed during a spell as a Leverhulme Visiting Professor in the United Kingdom, Steve is also pleased to acknowledge the support of the Leverhulme Trust (grant number: H5434400-VP2-2017-014), the University of Reading, the University of Exeter, and the Uehiro Centre for Practical Ethics at Oxford University. On a more personal note, Steve sends love to Lynn, Ben, Matthew, Sophia, Xena, and Maggie, who held our small world together even while the larger one was in turmoil.

Arthur offers heartfelt thanks to Colleen Hayes and Shane Hubler, whose careful reading and editing of an early draft of this text were more helpful than they know; and, to Cody Dout: our real-world dialogues on racial justice and climate change inspired and shaped many aspects of those enclosed.

Finally, we both extend appreciation of our families and friends who so graciously put up with our distance as we devoted so much time and energy to completing this work.

Introduction

1. Approach

We write this book as philosophers who share a deep concern for the fabric of life on Earth at a time when it is becoming ever more frayed.

Steve, a moral and political philosopher by training, has been a pioneer in the young field of climate ethics since the nineties and much of his professional life has been consumed by it. Largely setting aside other philosophical projects, he has spent the last couple of decades engaging with what he sees as the fundamental ethical challenge of our time.

Arthur, currently working on his doctorate in environmental values, came of age near the turn of the millennium, hiking Wisconsin's native woodlands and paddling its wild rivers. He has lived his whole life in the shadow of climate change, a force that has brought not only environmental crisis but a conceptual crisis for environmentalists. He is dedicating his scholarship to addressing both.

This book is an introduction. We have written it in dialogue form as an active discussion between diverse perspectives. We intend it as an invitation, and aid, to deeper conversations, not as any kind of final word. Accordingly, we invite everyone – students, teachers, the powerful, the marginalized, young people, the retired – to engage with this crucial topic. We believe that all of us have a stake in our planet's future.

We want to acknowledge that people come to climate change from different positions and perspectives. Helpfully, the dialogue form allows us to highlight diverse stakeholders in the climate problem. Nevertheless, we have not included – could not include – every perspective and every worthwhile point. Our guiding thought has been to structure the book in a way that reflects the kinds of conversations we see going on now, in the places we are most familiar with, as a way of enriching those conversations and providing an entry-point into wider discussion. Yet we are keenly aware that there are multiple conversations going on all over the world, and what we see is only a small fraction of the picture. So, inevitably, we will have missed issues that are important and voices that should be heard. Moreover, thinking about climate justice is not a static enterprise; it has evolved over time and will no doubt continue to do so. For all these reasons, this book – like all books – is bound by time and place. That being said, we apologize in advance for our omissions, and for any errors of judgment, or emphasis, in our choices of what to include and how to include it.

We should also emphasize that the books in this series (Routledge's Philosophical Dialogues on Contemporary Problems) do not set out to provide a neutral take on their subject. Instead, the aim is to guide the reader through a way of reasoning about a topic, one which inevitably reflects the views of the authors, and so constitutes an opinionated introduction. Given this, though we aim to give voice to a number of diverse perspectives, and often move swiftly on from some points or topics, we do not necessarily expect our readers to do the same. Indeed, sometimes we are implicitly encouraging further engagement through emphasizing disagreement, posing stark questions, or having some characters make provocative claims. Hopefully, the chapters provide ample routes into interesting conversations that can go in many different directions, whether in the classroom or around the kitchen table.

Some of our own views appear at various points in the dialogues, and of course, our general outlook also influences how the book is constructed and how the conversations go (as with every book). However, we want to emphasize that while we place the character Hope at the center of the discussion, she is not a mouthpiece for our views. Hope is not *us*, nor are any of the other characters. Indeed, we disagree with Hope's perspective sometimes, and often put concerns we think should be taken seriously into the mouths of others. Moreover, our own views are not identical, nor are they always fully formed. We do not see any of these things as a failing. In this book, we set out to emphasize questions as much as answers.

Some of our choices may surprise readers. For example, we do not begin this book by defending the scientific basis of the threat of dangerous climate change from those who consider the problem overblown or non-existent. This is partly because we think it more productive to make a positive case first, which then puts discussion of various skepticisms in context. It is also because we are of the opinion that certain forms of skepticism – such as deep skepticism towards consensus climate science – set up a discussion for failure. As philosophers, we tend to think that extreme, thoroughgoing skepticism can never be completely refuted, whether one is talking about climate change, gravity, the existence of an external reality, or anything else. However, in our view that shows that there is something wrong with extreme skepticism, rather than with gravity, reality, or climate science. Given this, we decided that beginning with climate skepticism would start the book off on the wrong foot. Nevertheless, we agree that climate skepticism remains a perspective that needs to be engaged. Accordingly, we dedicate Chapter 2 to discussing various forms of it. Thus, although we encourage readers to begin with Chapter 1, any who can't wait to get into such questions should feel free to reverse the order, reading Chapter 2 before returning to Chapter 1.

2. Outline

Some readers like to have a sense of where they are going, so we will now give a brief overview of the volume. (*Spoiler alert*: those who – like us – prefer to let things unfold, should skip over this section.)

Chapter 1 is set not long after the election of former President Donald Trump. It introduces our main character, Hope, who is just entering law school. The chapter describes her discussion with fellow students about climate science, the state of international politics, and why climate change has proven so difficult to address. In thinking through this last subject, the characters begin by discussing the influential view that climate change is a *tragedy of the commons* and how this tragedy may manifest in local and international spheres. Then, Hope presents the rival *perfect moral storm* analysis, which aims to show how the tragedy of the commons approach obscures central concerns of justice by ignoring the critical global, intergenerational, ecological, and theoretical challenges embedded in the climate problem.

Chapter 2 takes a step back and considers whether the moral challenge of climate change is really as formidable as Chapter 1 assumes. The chapter covers an encounter between Hope and her father, who is skeptical about the importance of climate change. They first discuss the epistemological question of whether we are justified in believing that climate change is occurring and human-caused, and the ethical question of whether some kinds of scientific skepticism are irresponsible. Their conversation then turns to skepticism about the role of ethics in climate action, where they discuss national self-interest, technocratic policy solutions, and the threat of climate extortion.

Chapter 3 investigates what role individuals should play, ethically speaking, in addressing the climate problem. Some of the characters from Chapter 1 return, tasked with compiling a list of climate recommendations for their university. Tracing debates that have become common in everyday environmental discourse as well as in philosophical circles, the characters immediately run into disagreement regarding whether individuals can be morally expected to reduce personal emissions. One of the interlocutors, Eliza, insists that a focus on individuals' private lives is condescending, elitist, and misses the point. The other two push back, and this begins a sustained discussion about individual climate responsibilities that spans many subtopics, such as the potential harm of personal emissions, the nature of collective responsibility, and the ethics of carbon offsetting.

Chapter 4 moves forward a little in time. Hope, now a junior climate negotiator, calls a meeting with her counterparts from other countries in the hope of making progress that has so far eluded their more senior colleagues. Their discussion explores justice in target-setting and burden-sharing, and is relevant to discussions of mitigation, adaptation, and unprevented loss and damage. The chapter also emphasizes how one's social and political context – whether it is where one lives, one's race, one's gender, one's culture, one's experience, or one's wealth (e.g., one's positionality) – informs one's views on the climate problem. This speaks to the importance of both *procedural* and *recognitional* justice, including the need to incorporate a diversity of people and worldviews in the process of designing climate policy.[1] Despite Hope's best efforts, the meeting ends in an impasse.

Chapter 5 takes place shortly afterwards. Richard, another junior U.S. negotiator, has called a meeting to discuss an alternative way forward. He presents geoengineering as a possible strategy for avoiding the many controversies embedded in mitigation, adaptation, and loss and damage. With the help of technical experts he has invited along, Richard describes two different kinds of geoengineering technology: Carbon Dioxide Removal and Solar Radiation Management. The negotiators discuss both the technical and ethical concerns surrounding research into, and the possible large-scale deployment of, both types of technology. For example, they deliberate about whether pursuing geoengineering might distract from mitigation efforts, and what the political implications might be. They then move on to consider deeper issues such as the ethics of actively intending to change the climate, the question of "playing god," whether geoengineering is "the lesser evil," recognition of nonanthropocentric worldviews, and the alleged advent of a new geological era, the Anthropocene. After extensive discussion, the negotiators again reach an impasse, and Richard departs with a pledge to hold another geoengineering roundtable restricted to interested parties.

1 See also Chapter 5.

Chapter 6 takes place sometime in the 2060s in a climate future where average global temperature has now increased by more than 2 degrees Celsius above pre-industrial levels, and continues to rise. In this future, record-setting heatwaves scorch the summers, many are suffering, and some desperately look towards geoengineering for solutions. Hope, now an older woman, discusses the failings of her generation with her granddaughter Caitlyn on the porch of their home in Washington, D.C. The chapter recounts Hope's words as she tries to describe some of the ethical challenges her generations faced, including how to put economic value on future well-being, how to understand our obligations to future people in the face of the non-identity problem, and whether the profound and unprecedented challenge of climate change required entirely new ethical concepts that her generations simply lacked. Caitlyn considers her grandmother's explanations, but finds herself frequently unsettled and dissatisfied. They close their discussion by considering how hope, courage, and other environmental virtues are still needed, and how a stable, beautiful, and just future may yet come about. This ending invites the reader to think about what such virtues might look like in our time, and how they may enable us to realize a future better than Hope and Eliza's.

3. Advice for Instructors

This volume can be read simply as a normal book and is aimed at a wide audience. However, it is also intended for classroom use, and so we are including some advice for instructors. The book can support various ways of structuring an introductory unit on climate change or climate ethics. We have trialed much of it in a 200-level introduction to environmental ethics course, and a 100-level contemporary moral problems course. In both cases, we were encouraged to see that most students preferred the dialogues to traditional textbooks. Our hypothesis, based on student testimony and our own reflection, is that this is due to the ways in which the conversational format makes the topics more accessible, memorable, and real.

One question that comes up is how much content to cover and how quickly. Here, the limiting factor isn't the amount of reading time. We've observed that students find it comfortable to read one dialogue to prepare for a particular class session, even though each dialogue is longer than a typical academic paper. Probably, this comfort is due to the format both being more engaging than most textbooks or academic articles, and facilitating relatively quick reading comprehension.

More significant limiting factors are the instructor's aspirations in covering each topic in whatever depth is appropriate to the course, and in providing the students sufficient opportunity to discuss the main ideas. Notably, each dialogue covers substantial ground. On the one hand, if one is happy to move slowly, each dialogue could provide enough material for a couple of days of introductory lectures, possibly supplemented with an additional reading (see annotated bibliography for some suggestions). We have divided each dialogue into two parts of roughly equal length to facilitate this "in-depth" strategy. We have also found that a single chapter can support discussion in a more advanced, seminar-style class, supplemented with a more traditional and focused reading.

On the other hand, it is feasible to cover most chapters in only one class session, so long as one is willing to be selective, and focus in on specific ideas and arguments. For this strategy, it is worth highlighting your emphasis to the students in advance. The chapters are divided into subsections to facilitate this. However, we would not recommend assigning only specific subsections, since the narrative within each dialogue matters.

A second question is how much of the book to teach. Obviously, one option is to read the entire volume, and we believe that the book reads best when presented as a coherent whole. As such, it would provide a good basis for, say, six weeks of classes (e.g., twelve lectures), probably supplemented with other readings. Hence, for example, one might use it as the basis for a substantial portion of an introductory course in environmental ethics, or global justice, or applied ethics, or environmental studies, or sustainability and justice, to name just a few possibilities.

That said, each chapter is more-or-less freestanding, so instructors can be more selective. For example, some courses on global justice may want to forgo the chapter on individual responsibility, while those teaching an introduction to philosophy course may be less interested in the chapter on technological fixes. Similarly, some courses may find some sequences of chapters or individual dialogues especially useful. So, for example, introductory ethics courses may find Chapters 3 and 4 helpful; general introductions to philosophy may favor Chapters 2 and 3; while courses in global justice or international politics may choose Chapters 4 and 5. If teaching multiple chapters, we encourage doing so in chronological order, since the overarching narrative of the dialogues makes reordering topics more awkward than it would be with a traditional textbook.

A third question is how to make best use of the dialogue format in teaching. One issue for some philosophers might be that the dialogues render the philosophical discourse between characters somewhat less analytic than you would find in a conventional textbook or academic article. For instance, you will not find premise-by-premise argument reconstructions within these pages; thus, students and instructors may have to do a little more philosophical work themselves. We tend to think that this is not a bad thing. Not only does the accessibility of the dialogue format make this sacrifice worthwhile, but the intellectual challenge also mirrors those students will face in their everyday lives.

More importantly, perhaps, the dialogue format encourages pedagogical innovation. Here, we list a number of recommendations that we have found successful in the past, or plan to implement in future classes.

Paired Readings

As we have noted, since this book makes for easy and relatively painless reading for students, it is not unrealistic to assign other readings to accompany the chapters. Which readings you choose is, of course, going to be a matter of taste, and will depend on what topics you wish to emphasize. One approach is to use academic articles to explore some questions in more depth. For inspiration, we encourage you to look at the annotated bibliography, and the primary literature we reference liberally throughout. Another strategy, perhaps of special appeal in interdisciplinary courses, would be to intersperse op-eds, policy briefs, interviews, or other short, public-facing pieces.

Emphasize Examples & Eye-opening Ideas

An overarching goal of this book is to convey the central concepts of climate justice in a way that feels more grounded, applicable, and interconnected than they often are when taught in classrooms. To this end, we have incorporated timely examples and provocative concepts throughout the chapters. We encourage instructors to emphasize these examples in their lectures, and in whatever activities they decide to pair with their lessons. The following is a list of some particularly useful examples in our judgment:

- Chapter 1, part 1: The Right to be Cold (p. 9; relevant discussion on pp. 23–24)
- Chapter 1, part 2: The Tyranny of Humanity (pp. 36–39)
- Chapter 2, part 1: Contrasting Standards of Evidence in Law (p. 63; relevant discussion on pp. 63–68)
- Chapter 2, part 2: "Pittsburgh, Not Paris" and National Self-Interest (p. 69; relevant discussion on pp. 69–72)
- Chapter 3, part 1: Joy-guzzling (p. 105; relevant discussion on pp. 103–105)
- Chapter 3, part 2: Comparing Lives: Greta Thunberg and Luisa Neubauer (pp. 117–119; relevant discussion on pp. 113–120)
- Chapter 4, part 1: 2 Degrees: A Suicide Pact for Africa (pp. 159–160; relevant discussion on pp. 157–160)
- Chapter 4, part 2: The Great Wall of Lagos (pp. 166–168)
- Chapter 5, part 1: Assuming Negative Emissions: Unjust Gambling in IPCC Modeling (pp. 198–200)

- Chapter 5, part 2: Indigenous Worldviews (pp. 217–220)
- Chapter 6, part 1: Discounting for Pure Time Preference (pp. 239)
- Chapter 6, part 2: Radical Hope (pp. 270–271; relevant discussion on pp. 269–272)

Tailor Assignments to Dialogues

This text encourages novel assignments, and we encourage experimentation. Below are four possible assignments we suggest that take advantage of the dialogue format.

Perspective Presentations

Each student (or group of students) takes a view articulated in a given chapter or topic, and develops the perspective further. Perspectives can either be assigned by the instructor or picked by the students. Students should follow the relevant footnoted references in the book and track down the primary citations. Then, they give a short presentation on their research (5–15 minutes, depending on course). This assignment helps students develop research skills, and reflects the spirit of the book by further highlighting the nuanced and diverse perspectives present in climate discussions.

In-Class Dialogues

Pick a part of one of the dialogues, and assign each student or group of students a character. Students spend 5–10 minutes or so coming to a consensus on the perspective of their character. The instructor then picks a topic of discussion, possibly taken from the chapter subsections. The students are given time to decide what position on the topic their character would argue for and how (e.g., 10–15 minutes). They then present their arguments to the rest of the class (e.g., 5 minutes per group). Each group should take notes on the key points made by other groups during their presentations.

Then, the class can move on to free-form negotiation moderated by the instructor (e.g., 15 minutes). During this time, the students see if they can come to a compromise on the topic at hand that is within the values of their character.

Note: This activity works best for chapters involving three or more characters (e.g., consider students taking on the perspectives of country representatives in Chapter 4 and 5, or the perspectives of the three characters in Chapter 3).

What Do You Think?

In this written assignment, students pick a place of disagreement between two characters in the dialogues. They then reconstruct the argument made by one of the interlocuters, and an objection made by the other character. After doing so, the student explains their own position on the disagreement. This essay assignment is similar to a typical argumentative philosophy paper, except the students have the extra challenge of expanding on a disagreement that may have gone by (relatively) quickly in the dialogue compared to a disagreement between professional academics as it played out in a journal. To this end, it may help students to follow the relevant footnoted references. This assignment teaches the classic philosophical virtues of precision and being charitable to the views of others while also requiring them to fill in unstated premises in an interlocuters' argumentation, in ways similar to normal interpersonal contexts.[2]

Create Your Own Dialogue

In this written assignment, students write their own dialogues on a topic related to climate ethics or climate justice. The dialogues

2   You might require the students to state their own position in the following fashion. If they agreed with the position of the argument initially reconstructed, the student gives an original argument as to why they think the objection fails. If they agreed with the objection, the student presents a possible reply on behalf of the original arguer and then gives an original argument as to why this reply fails.

should involve at least two characters with meaningfully different perspectives, with a minimum word length of the instructor's choosing. This is an assignment where student creativity should be encouraged. For a fun twist, students can partner up and then perform the dialogues for the class.

Integrate Active Learning

The dialogues encourage active learning that is engaged in ongoing conversations and contemporary issues. Ideally, assignments would not be confined to the classroom. We encourage the instructor to integrate active learning. Below are five possible ways to do so, but this is a place where innovation and experimentation are very good things.

Society in a Warming World

Students (independently or in small groups) find a news article or op-ed outside of the course related to recent climate events, and give a report connecting it in some way to class discussion.

Changing Environments in a Warming World

Students are asked to go out into a natural area, or a relatively wild part of the urban environment, and to spend half an hour attuning to the sounds, smells, sights, and general character of the surroundings. Then, the students should reflect on the experience in journal, video-log, or presentation form, and posit how they think such an environment may change in a warmer world. Their reflection need not emphasize the negative changes they anticipate, but might imagine positive changes that may come as well.

Family in a Warming World

Katharine Hayhoe, an atmospheric scientist and climate communicator, says that one of the most important things you can do about climate change is talk about it.[3] In this spirit, students talk to an older family member about their views on climate change: whether they think it's a problem, what they think should be done, and whether they think ethics or justice is important to consider when thinking about the problem. The student should prepare questions to ask their family member beforehand, to reference if need be. These questions can be submitted to the instructor. The student then reflects in written or presentation form about how this conversation went. This assignment should ideally be completed after all or most of Chapters 2, 3, 4, and 5 have been read, but before reading Chapter 6.

The Individual in a Warming World

Students are asked to keep a journal throughout the class on how they see themselves in relation to climate change. For instance, after Chapter 3, the student might reflect on how the class did or did not change their thinking about own behavior: their personal emissions, their collective action, or anything else.

Hope in a Warming World

A recent survey of over 10,000 young people (aged 16–25 years) found pronounced levels of climate anxiety associated with perceptions of inadequate action by adults and governments, feelings of betrayal, abandonment and moral injury.[4] In an active learning activity done during class time – preferably in a comfortable atmosphere outside – students discuss their own feelings related to climate change and share any strategies they use for staying hopeful.

Of course, these are just a few ideas for pedagogy. We encourage you to teach this text however you best see fit, and hope that the dialogue format might encourage and facilitate innovation. Once the book is out, we'd love to hear suggestions from other instructors too!

3 Hayhoe, 2018.
4 Marks, *et al.*, 2021.

Dialogue 1   Why Ethics?

Fall 2017: shortly after President Trump's withdrawal from the Paris Climate Agreement

Part 1: An Introduction to the Climate Change Problem

Butterflies flutter in Hope's chest as she approaches the door to the conference room. She is a little early, as she always is when she is nervous or excited. Surprisingly, she feels more apprehensive than she was when she defended her master's thesis just a few months ago. The stakes were probably higher then, but she knew her committee well. Here, at one of the best environmental law programs in the country, she knows no one. But that was one of the reasons she wanted to join the campus climate activist group in the first place, and why she agreed to be on the steering committee. Plus, this is the perfect opportunity to apply her skills and make a real difference. Today, she begins her fight for climate justice in earnest.

Hope pauses to take a deep breath before entering the room. To her surprise, someone is already there. A young woman around Hope's age, smartly dressed and studying a tablet with rapt attention – notes, presumably. Hope feels suddenly underprepared. The woman looks up and smiles, and Hope's anxiety vanishes.

Hope: Hello. I'm Hope, the ethicist.
Eliza: Hi, I'm Eliza. The ethicist?

DOI: 10.4324/9781003123408-1

Hope: I'm a new law student, but I've just finished a Master's degree in philosophy, working on climate ethics. I'm especially interested in issues of justice and responsibility.

Eliza: (*smiles*) All sounds very worthy. But why are you *here*?

Before Hope can answer, they are interrupted by the entrance of four others.

Stacy: Hiya, I'm Stacy, the scientist. This is Andrew. He's an activist from Climate Justice Now. He's brought along his friend, Adama, from the Anti-Colonial Alliance. I see you've already met Eliza, our economics and policy person. And this is Nadia, my best friend.

Nadia: I'm the designated nonexpert! The normal one. (*smiles*) I don't know much about climate yet, but I'd like to. I'm your "sounding board." I'll stop you when you get too technical – or start sounding pompous!

Hope: Nice to meet you all in person! Thanks for putting this together, Stacy.

Introductory Climate Science

Stacy: Of course! Well, we're all busy and we don't have a ton of time, so let's get started. As you know, the point of this meeting is to plan our first pitch for climate action to the other students. My background is in environmental science. I thought I'd cover climate science, then we'll have Eliza talk about governance and policy, Hope about ethics, and Andrew about activism. So, I'll go first, laying out the basics. You know: the idea that the Earth's climate is already changing due to human activities, especially the burning of fossil fuels, and that if we don't change our behavior quickly then over the next century we will see dramatic shifts that will threaten human life, global civilization, and the nonhuman world. Massive suffering, death – you know, scary stuff like that.

Nadia: I think you'd better be more specific.

Stacy: Already with the interruptions! But you're right, of course. The way that I like to introduce it is by pointing out that the Earth's atmosphere plays a vital role in controlling the climate.

Nadia: You mean that without the atmosphere, the earth would be much colder?

Stacy: Yes. In fact, the average temperature would be much *too* cold for life as we know it to exist. The surface is only as warm as it is because the layer of gases that constitute our atmosphere acts like a blanket trapping energy from the sun. Kind of like a greenhouse. So, being as imaginative as ever ... (*Stacy winks*) ... scientists call these gases "greenhouse gases," and the overall phenomenon "the greenhouse effect."

Nadia: So, greenhouse gases are actually good?

Stacy: Up to a point. Without the natural greenhouse effect, we'd basically freeze to death. But beware too much of a good thing, as they say. Since the onset of the industrial revolution, humans have been adding large quantities of greenhouse gases to the atmosphere and increasing the concentration. A major source is the burning of fossil fuels, which releases carbon dioxide. Before the industrial revolution, the atmospheric concentration of carbon dioxide was about 270 parts per million.

Nadia: What is it now?

Stacy: Over 410 parts per million. Most of the emissions have occurred in the last sixty years.[1] By the end of the century it could be anywhere between 400 and 1100 parts per million.[2]

Nadia: Wow! That's a big range. Are scientists really that uncertain?

1 Intergovernmental Panel on Climate Change (IPCC), 2021b, p. 5. See annotated bibliography.
2 IPCC, 2021a, p. 47.

Stacy: Well, most of the uncertainty involves how humans will behave moving forward, not the background science. If we aggressively reduce our emissions, we can keep the carbon concentration to around 400. If we don't, or – God forbid – if we *significantly increase* emissions ....[3]

Nadia: It'll be way higher.

Stacy: Yes. Either way, we're making a big change to the chemistry of the atmosphere. Scientists have modeled these changes. They project where we're likely headed, based on theories of how the climate system works and data about past climate systems.

Nadia: What do these projections say?

Stacy: Let's take the global surface temperature late in this century, averaged over 2081–2100. For a very low emission scenario, the temperature rise is estimated to be 1.0°C to 1.8°C; for intermediate scenarios 2.1°C to 3.5°C; and for a very high emission scenario, 3.3°C to 5.7°C.[4]

Nadia: Is that even a big deal? It doesn't sound like it. If it were 2 or 3 degrees warmer tomorrow, I might not even notice the difference! Probably 4–5 degrees wouldn't make that much difference to me either. So, why does it matter if it is a few degrees warmer in 2100?

Stacy: Unfortunately, this is a place where our normal ideas about the weather mislead us. Crucially, shifts in daily temperatures are not a good comparison class for shifts in global average temperatures.

Nadia: Really?

Stacy: Yes! During the last ice age, there was a kilometer of ice – sometimes miles of it! – over Northern Europe, and places like New York, Seattle, and Boston. How much colder do you think it was on average globally?

Nadia: I've got no idea, but it must have been a massive temperature shift. 20 degrees colder? More?

3   Wuebbles, *et al.*, 2017, p. 151.
4   IPCC, 2021b, p. 17.

Stacy: On average, it was 4.5 degrees Celsius colder than the twentieth century norm.[5] In general, ice ages are only 5–7 degrees colder on average than the warmer interglacial periods.[6]

Nadia: Wow! So, you're saying that climate change might be as big as an ice age shift, but in the warmer direction? Actually, even half an ice age shift sounds pretty major …

Stacy: Exactly. Dramatic stuff. Also, that is not the end of the story. My numbers were for late this century. But why fixate on that? If we look further out – beyond 2100 – the numbers often get higher. So, for example, when we make projections to the end of the twenty-third century, even the IPCC's intermediate scenario increases to an estimated 2.3–4.6°C rise, while a very high emission scenario goes to 6.6–14.1°C.[7] If we burned all known reserves of fossil fuels, we might ultimately see average warming of 6.4–9.5°C.[8]

Nadia: I'm in shock. What could that magnitude of change possibly look like?

Stacy: It is really striking. Suppose we ask about an 8 degree rise in the global average. The last time the world was that warm it was a very different place. There were crocodiles at the Poles![9] Worse, this time, the shift would be very fast. It might occur over a number of decades, or a couple of centuries, not over thousands of years, as happens in an ice age shift.

---

5   For a visual representation, see https://xkcd.com/1379/.
6   The global average conceals regional variations. For instance, some middle and high latitude regions of the Northern Hemisphere saw changes of 10–15°C during the ice age transitions. Similarly, moving forward, warming may be much higher in some regions. For example, Arctic temperatures are projected to increase at two times the rate of the global average (IPCC 2021a, SPM 19).
7   IPCC, 2021a, 2–10.
8   Tokarska, *et al.*, 2016.
9   Lavelle, 2016.

Nadia: So, nothing at all like the weather being just a little warmer tomorrow.

Stacy: Right.

Nadia: Hmm. Is this why you think it's not a good objection to climate science to say that the world has been hotter in the past? I know that objection bugs you.

Stacy: Right. For one thing, humans weren't around when it was that much warmer. More importantly, I doubt that human civilization could adapt to such dramatic change happening so quickly.

Nadia: You mean, like Canada suddenly turning tropical and large parts of America into desert, that kind of thing?

Stacy: Yes. Still, you don't need to focus on the very high end to get really concerned. For instance, some believe global agriculture may collapse if the global average temperature rises even by 4°C.[10] So, at that point, we're already talking about potentially fracturing the fundamental preconditions of human life and society.[11] A "crash landing" scenario, if you will.[12]

Nadia: (*gasps*) I had no idea.

Stacy: In any case, even if it is difficult to know exactly what would happen, when, and where with these kinds of increases, it seems clear that we are talking about very major shifts.

Nadia: How strong is the evidence for all this?

Stacy: Pretty strong. The main mechanism is basic science and has been known for a long time. Other things being equal, if we increase greenhouse gases, the planet will warm. We understand that. Over the last century, scientists have recorded temperature rise worldwide. This rise

10 Weisbach, 2016, 178, citing New, *et al.*, 2011.
11 Anderson, 2012; Moses 2020, quoting climate experts Will Steffen and Hans Joachim Schellnhuber; Lynas 2020; see Gardiner 2011a, 159.
12 Gardiner 2011a, chapters 6–7. Part of what is relevant here is the possibility of "tipping points" in the climate system. See Lenton et al. 2019.

is consistent with our best predictions. We would expect about a degree by now, and that's what we're seeing. I've got lots of data and graphs. I'll put up a couple in the presentation.[13]

Nadia: What about those who think climate change isn't happening? The skeptics, I mean.

Stacy: Well, there isn't really that much skepticism among the scientists, especially about things that matter for the kinds of action we need to take now. As I said, the basic mechanism is well known, and observations over the last few decades are consistent with what scientists would expect. For a while, there were outstanding questions about whether other mechanisms might be in play that might offset or dampen down the human influence. But scientists have been looking at those for decades and haven't found anything that makes a big difference. For example, we know that it's not sunspots, or other natural variations causing most of the current temperature rise.[14]

Nadia: (*interested*) Really?

Stacy: Yeah. The impact of increasing greenhouse gases swamps all the other mechanisms that might make a difference. Also, the current temperature rise is only the beginning of what we're likely to see. So far, the global average is up by about 1.1 degrees Celsius over pre-industrial averages, but the most aggressive plans internationally only call for limiting it to 2.0°, or *maybe* 1.5°. Temperature will keep rising until at least the mid-century with near certainty, and absent deep reductions in emissions in the coming decades we'll fly past the 2-degree threshold.[15] So, there is more warming to come.

Nadia: How soon can we expect significant impacts?

13 See IPCC 2021a.
14 Wuebbles, *et al.*, 2017, pp. 36–37.
15 IPCC, 2021b, p. 17.

Stacy: Well, we're already seeing rapid change in Alaska, and even in the lower-48 states of the United States, we seem to be experiencing some climate effects. Heatwaves have become more frequent, and extreme cold less so. The incidence of large forest fires has increased. We're seeing earlier spring melt and reduced snowpack already affecting water resources in the Western States. Climate change has also probably worsened extreme weather. We have good reason to think it contributed to the devastating effects of Hurricane Katrina, for instance. Of course, these are just the early signs of a problem that will continue to get worse – potentially unimaginably worse if we don't act. More extreme wildfires, hurricanes, and droughts fueled by climate change *will* end lives and disrupt livelihoods, here in America and around the world. Within the next hundred years, we can also expect significant sea-level rise, threatening coastal cities like San Francisco and New York.[16] *Even under low emissions scenarios*, the projection is for 1–3 meters sea-level rise within the next few hundred years.[17] Three meters is nearly 10 feet! If that happened, the United States would lose 28,800 square miles of land, which is home today to about 12.3 million people.[18]

Hope: Well, I've got members of my own family who think they know better. They insist that climate changes naturally, and that its silly doomsday talk to suggest we could not adapt to whatever change happens.

Stacy: (*sighing*) Most of us have family and friends with those kinds of views. I plan to let that come out in the question and answer, if it comes out at all. I don't like highlighting it at the beginning, before we even get the basic scientific picture out there. I think doing so tends to make

---

16 Wuebbles, *et al.*, 2017, p. 18.
17 IPCC, 2021a, p. TS-45.
18 Strauss, 2014.

skepticism look like more of a serious issue for climate science than it really is. With the basic picture in hand, it is much easier to talk about any specific questions people might have.[19]

Nadia: Okay. If the science is pretty robust, why do you think some remain unconvinced?

Stacy: I'm sure there are many reasons. One is probably that the warming climate does not affect all of us equally. For instance, denying climate change may seem more plausible for many well-off Americans when they either don't directly experience the effects yet, or those effects seem manageable to them. By contrast, some people and cultures are already feeling the threat in their bones, and it is an existential threat. For example, it is warming more quickly in the Arctic than anywhere else, and Inuit people who live across Canada, Alaska, Greenland, and Russia rely on their ice and snow for not only their *physical* survival but also their *cultural* survival.

Hope: I'll never forget learning this from Sheila Watt-Cloutier, an Inuit activist. Snow serves as highway for their dog-sleds; ice serves as home for the creatures they hunt for food and celebrate in their stories. Even back in 2010, Watt-Cloutier was saying that rapid, tumultuous change has come to her homeland in her lifetime, to the extent that the only option left for some communities was likely costly relocation. She argued that it was important to see climate in terms of threats to human rights, including rights to health, physical security, subsistence, and culture. It really struck me when she said, "as our culture is based on the cold, the ice and the snow, we are in essence defending our right to be cold."[20]

Nadia: Wow. I didn't realize that the problem is already that bad for some people.

---

19 For discussion of climate skepticism, see Dialogue 2.
20 Watt-Cloutier, 2010. See annotated bibliography.

There's a pause. The students exchange glances, eyes serious.

Climate Politics

Eliza: I suppose this is where I come in?
Stacy: Sure, I'll stop there and leave it to you to talk about the policy and politics.
Eliza: Got it. The good news and the bad news?
Andrew: Is there any good news?
Eliza: Not much, but they need to know about that too.
Andrew: Absolutely.
Eliza: I thought I'd begin by telling them about the United Nations Framework Convention on Climate Change.
Nadia: That's a mouthful!
Eliza: (*laughs*) True! People say U-N-F-triple-C for short. The UNFCCC is the foundational treaty on climate change, forged at the Rio Earth Summit in 1992.
Nadia: What was that all about? I've heard of it, but what did it say?
Eliza: It was a pretty big deal. The countries of the world got together and agreed to prevent "dangerous anthropogenic interference with the climate system." They also announced some principles and guidelines for doing that, including that developed countries should take the lead.
Nadia: Anthropogenic?
Eliza: (*grins apologetically*) That's just a fancy name for "human-caused." Basically, the world agreed that climate change was a serious threat, that human civilization was causing it, and that all countries needed to come together to confront the problem.
Nadia: They really all agreed?
Eliza: Everyone signed up. All the major greenhouse gas emitters, including all the major powers, and almost every other country besides. So, the United States, Russia, China, the European Union, Brazil, India ....

Nadia: Sounds awesome! But ... wait, that was thirty years ago. If everyone agreed, shouldn't climate change be under control by now? Why are we still talking about this?

Andrew: Sadly, the climate situation is way out of control. Annual emissions of carbon dioxide are up by more than 40% globally. When you factor in other greenhouse gases, the atmosphere's heat-trapping capacity has gone up by around 60% since 1990.[21] Right, Stacy?

Stacy nods, face grim.

Nadia: What went wrong?

Eliza: It looks like some combination of procrastination, low ambition, and outright bad behavior. Over the years, countries kept making agreements that sounded good in principle, but were weak and inadequate in practice. Within a few years, even those weak agreements would be sabotaged by one or two major actors. Usually, it was the United States backing out.

Nadia: Oh, no!

Eliza: Yep, sadly. In Kyoto in 1998, 160 nations, including the United States, agreed that the industrialized countries would cut their emissions 5% by 2008 to 2012. However, in 2001 President George W. Bush announced that the United States was withdrawing. In Copenhagen in 2009, countries agreed to limit temperature rises to 2 degrees Celsius, but they did not make the commitments needed to back that up. Then, in Paris in 2015, all 196 countries – including the United States, which led the effort – agreed to cut their emissions to meet the 2-degree ceiling and aim for closer to 1.5 degrees C.[22]

Nadia: But the Trump administration backed out of Paris, right?

---

21 NOAA, 2020.
22 For discussion by a philosopher engaged in climate negotiations, see Light 2016.

Eliza: (*sighs*) Yes. Just this summer [summer 2017], President Trump declared that the United States would withdraw in 2020 – the earliest opportunity available – *and* not implement its previous commitments.

Stacy: I'm praying the next president will rejoin.[23]

Hope: I hope so. Still, I'm worried about this history of grand agreements that promise a lot, but don't seem to accomplish much of anything. That global rise of 60% after thirty years looks like a big failure to me.

Eliza nods sadly.

The Tragedy of the Commons

Nadia: What caused the failure?

Eliza: There are lots of potential explanations, and probably some truth in most of them. Probably the most popular is that climate change is a tragedy of the commons.

Andrew: You mean like Hardin's famous example?[24]

Nadia: What's that?

Andrew: Garrett Hardin was an ecologist. He asked us to imagine a common pasture, like the traditional English commons. You know, a plot of land in the middle of the village where everyone has the right to graze their own animals.

Nadia: Sounds idyllic!

Andrew: Well, Hardin's idea was that it actually leads to disaster! (*chuckles*) His example is supposed to explain why environmental degradation of common land occurs. The story goes like this. Each cattle owner – Hardin calls them 'herdsmen' – wants to add more animals to his herd,

---

23 President Joe Biden rejoined the Paris agreement on the first day of his presidency (January 20, 2021) by executive order.

24 Hardin, 1968. See annotated bibliography. Hardin's main focus in the paper is actually on world population growth. For a detailed critique of his population argument, as well as a comparison to climate change, see Gardiner, 2001.

so he can sell them at market and increase his profits. But each additional animal also puts extra strain on the commons. It eats more grass, leaving less for the others. Unfortunately, these costs are likely to be ignored since there are problematic incentives. Whereas the individual herdsman gets the benefits from adding to his own herd, the costs – the environmental damage that results – are shared by all. So, Hardin says, it is predictable that – for the herdsman personally – his benefits will outweigh his costs, and make it rational for him to buy more animals even if that is bad overall for the commons.

Eliza: (*smiles ruefully*) But the real kicker is that the same is true for all the herdsmen. Everyone is in the same position. So, they assess the situation in the same way, and also add extra animals. But this means that everyone is adding extra animals. So, now we're talking about a lot of animals, not just a few. The upshot, then, is severe overgrazing. Ultimately, if all the herdsmen keep acting on their individual incentives, the commons is destroyed: no one can graze their animals, and all suffer.[25] Hardin says: "ruin is the destination toward which all men rush, each pursuing his own best interest in a society that believes in the freedom of the commons."[26]

Nadia: Okay; I see that. But isn't this just a normal case of thoughtlessly exploiting the environment? Why is this example so well known?

Andrew: It's famous because of the central paradox Hardin identified: the situation is set up so that when each person acts for their own benefit, they all arrive at an outcome that is not in anyone's interests, including their own.

25 The situation is actually a little more complicated. For example, usually, it is assumed both (i) that each herdsman adds animals only a few at a time, and (ii) that no single addition (and no single round of additions) is enough to destroy the commons immediately. Among other things, this suggests that the imagined tragedy evolves over time. See Gardiner, 2011a, ch. 4.
26 Hardin, 1968, p. 1244.

You might say that *individual rationality* leads them to destroy the commons, even though all agree that it would be *collectively rational* for them to preserve it.

Nadia: I agree that it sounds like a bad situation. Still, I'm not sure why it is supposed to be "individually rational" to add more animals. Surely, when the herdsmen see what's going on, they would just stop and not overgraze? It is not in their interest *in the long run*. That's our whole problem in this society! We don't think ahead!

Andrew: Believe it or not, the problem is not ultimately lack of knowledge.

Nadia: (*tilting her head in puzzlement*) What do you mean?

Andrew: Well, let's think about those incentives, and the decision that each herdsman has to make about whether or not to add more animals. He faces two possibilities: either the others will add more animals, or they won't. However, it turns out that *either way – regardless of what the others do* – the best thing from his individual point of view is to add more animals.

Nadia: Why do you say that?

Andrew: Well, consider the choices. On the one hand, let's say the herdsman believes the other herdsmen *will* add more animals, or at least guesses they will; then, the commons is going to be destroyed anyway. Given that, it still seems better for him to add more animals. After all, he will get something from doing that: the profit from whatever extra animals he can graze and sell before the commons collapses. If he holds off while everyone else is adding animals, then he loses even that. So, there's a kind of "race to the bottom" problem.

Nadia: Race to the bottom?

Eliza: That's an economics phrase! In this case, I suppose it means roughly that lack of regulation incentivizes a downward spiral of environmental exploitation. Right, Andrew?

Andrew: That about captures it.
Nadia: Hang on. You're just assuming that the others will not cooperate. But it is not surprising that this leads to bad outcomes – and for all of the herdsmen. What about if they do cooperate?
Andrew: Okay – that's the other side of the dilemma. Suppose the herdsman thinks that the others are *not* going to add more animals. Still, there's a strong incentive for him to add some. After all, the commons won't be destroyed by him alone adding just a few more animals. So, he can "free ride" on the cooperation of the others.
Nadia: That's unfair!
Andrew: True; but I think we can see why it is *tempting* for him to add another cow, when doing so helps him and his family.
Nadia: (*flustered*) I still think that would be wrong! Lots of bad behavior might benefit us or our families – lying, cheating, stealing, etc. – but that isn't a good enough excuse!
Hope: I agree. Nevertheless, it is interesting how quickly we turn to questions of fairness and ethics. It doesn't seem *merely* a problem of self-interest or economics anymore.

A brief silence ensues as the others digest this thought.

The Climate Commons

Nadia: So, how does this explain why countries haven't taken real action against climate change? That was the point of bringing up this example, right?
Eliza: Right. On the conventional story, the idea is that each country faces a conflict between collective rationality and individual rationality. Picture the atmosphere as a global commons, like the pasture. Of course, the atmosphere is a bit more complicated than a field! Still, just like the pasture, the atmosphere can degrade if its members do not constrain their actions, and everybody is worse off.

Nadia: Sure. That's what Stacey has shown us.

Eliza: So, let's assume that each country thinks that it would benefit it to avoid dangerous climate change. Then, they all accept the need to limit overall emissions. They agree that this is collectively rational.

Stacy: (*thoughtfully*) Wait. In theory at least, I suppose they could also increase their carbon sinks, right? For instance, they could plant trees or build machines to suck up CO_2 so as to offset the increase in emissions ...

Eliza: That's true. You're noticing already how the atmosphere is more complicated than a pasture. Still, let's set that aside for now, for simplicity's sake.[27] (*Stacy nods*) Okay; so, suppose each country accepts the *collective* need to limit overall emissions. Nevertheless, when each tries to decide what it is going to do as an individual country, it may face perverse incentives. Most notably, it may have a strong economic incentive to emit carbon at high levels at least in the short term.

Nadia: This is going to be like the herdsman's incentive to add extra animals, right?

Eliza: Yes! And we see the same dynamic playing out. On the one hand, suppose our country is deciding what to do. If the government thinks that other countries will cooperate by reducing their emissions dramatically, it might think that we can get away with free-riding, at least for a while.

Nadia: Ugh – again, that's awfully unfair.

Eliza: Sure, but highly enticing! On the other hand, if our government thinks that the other countries will not reduce their emissions, then it may think that reducing our own emissions is futile. Since dangerous climate change will come about anyway, it may conclude that we may as well take what we can get before catastrophe strikes.

27   See the discussion in Chapter 5.

Nadia: Aren't there any solutions to the tragedy of the commons?

Eliza: Usually. In fact, there are numerous ways out. Hardin himself suggests external sanctions: "mutual coercion, mutually agreed upon."[28] The thought is that if there is a robust regime that punishes those who don't cooperate, this will shift the incentives. If not cooperating will result in sanctions, then it may no longer be individually rational not to comply.

Nadia: Ah, I see the issue (*with an air of realization*). I'm guessing that mutual coercion is tricky in the international arena. After all, who has the power to punish the big emitters, like the United States, China, the European Union, and so on? There's no sheriff in town. Or if there is supposed to be a sheriff – like the United Nations – it doesn't have enough power to do the job.

Eliza: Exactly!

Nadia: So, what other solutions are there?

Eliza: Informal sanctions can work. Say that the herdsmen are members of the same community. They might not be willing to add animals if that is something that will upset their neighbors. Perhaps no one will talk to them at the village pub on a Friday night if they get greedy!

Nadia: (*laughing*) You wouldn't want that!

Eliza: No doubt! In fact, informal sanctions like these may have helped many commons work effectively historically, at the local level.[29]

Nadia: So, why not try that?

Eliza: Some propose that. There's great work on the conditions where it tends to work well. Elinor Ostrom won a Nobel Prize in economics for her contributions – actually, she's a personal hero of mine, as the first woman to win that award![30] Still, the conditions that enable an unregulated,

28 Hardin, 1968, p. 163.
29 Ostrom, 2009. See annotated bibliography.
30 Piurek, 2019.

sustainable commons don't seem to fit well for global, intergenerational problems like climate change.[31]

Andrew: Also, arguably, it is what has been tried already, in the UNFCCC, from Kyoto to Copenhagen to Paris. These agreements are basically voluntary. In Paris, each country is asked to set out what its own contribution will be to addressing the climate problem – it's "nationally determined contribution." Then, there is an idea that they will be assessed publicly on whether they live up to those commitments. Some call this a "name and shame" approach.

Nadia: I can think of some leaders and countries who seem pretty oblivious to shame!

Andrew: I would laugh if it weren't such a problem. Naming and shaming hasn't worked. Some people say a tougher approach, like trade sanctions, might fare better.[32] But no one seems able to agree to that, presumably because trade is such a contentious issue anyway. I suppose it might work if a genuine superpower or group of superpowers pushed it, but the superpowers are the main problem in the climate case.

Eliza: Some have suggested forming a "climate club" of motivated countries.[33] The advantage of being a member might be access to other countries' markets with low trade barriers.[34] To remain a member, countries would need to take adequate climate action as defined by the club leadership. This model is supposed to counteract the incentives to freeride: if you don't do your part, you lose out on the perks of the club.

Andrew: While maybe sensible in theory, climate clubs don't exist yet. That suggests there are serious barriers to setting

31 Gardiner, 2011a, pp. 116–118.
32 Barrett, 1997; Gardiner, 2004; Nordhaus, 2015.
33 Buchanan, 1965; Victor, 2011; Nordhaus, 2015.
34 Nordhaus, 2015, p. 1340.

them up, given how long we've been wrestling with climate change. There are also big issues of power and political legitimacy. Who decides? Won't the big countries bully the others?

Hope: In the meantime, what about ethics? Why not self-restraint? Suppose the herdsmen care enough about their community for its own sake to stop. They don't want to harm their family, friends, and neighbors. They see that the commons is vital to their shared way of life – they don't wish to play a part in its collapse! In short, suppose they reject the rather narrow, self-centered sense of being "rational" that the whole tragedy of the commons thing plays into.

Stacy: Yes! It all sounds pretty yucky to me. I don't think real people are that self-centered!

Eliza: I agree that a more expansive set of motivations could do it. Perhaps that does happen in close-knit communities. (*sighs*) Nevertheless, I worry when it comes to a genuinely global and intergenerational commons like the climate system. You'd probably need a global ethic – one that gets the whole of humanity to consider themselves a single community, to come together in kinship. Governments would need to really embrace that ethic for it to work at the international level; so would people and communities at the local and individual levels. I hope that humanity realizes such a world, but it seems a big ask right now, and we don't have much time.

Nadia: Hmm. Stepping back, I think I understand why so many consider climate change to be a tragedy of the commons. It seems to explain why we see the cycles of attempts to do something, followed by backsliding and failure. *Collectively*, countries know that global climate change is bad for them, but when national governments decide what to do *individually* they often decide to emit too much anyway, perhaps because they are focused only on a narrow set of interests. If this continues, it seems predictable that eventually the commons will collapse due to global warming.

Part 2: A Perfect Moral Storm

Hope: Not so fast. I agree that the tragedy of the commons analysis seems helpful. Still, I tend to think that matters are much worse, much more complicated, and more fundamentally a matter of ethics and justice.[35]

Nadia: (*aghast*) How could things possibly be *worse*?

Hope: Have you ever heard that story of "the perfect storm"? It's about a fishing boat caught at sea as several independently powerful storms converge on it.

Stacy: Sounds familiar! During my fieldwork last summer, two storms converged on our vessel, and we barely made it out. Each storm was pretty small, but together they created chaos.

Hope: That sounds scary enough. But in the "perfect storm" story, each storm *by itself* might well have been powerful enough to sink the boat. One can hardly imagine what it was like when they combined. Terrifyingly, the book is based on an actual event.[36]

Andrew: Spoiler alert. The boat sinks.

Nadia: (*swallows*) Sounds horrific. How does this relate to climate change, though?

Hope: Well, the idea is that, just like that boat, we're caught between several independently powerful and mutually reinforcing challenges to ethical action – or "storms." Each "storm" is strong enough taken individually that it might blow us off course, away from what we morally ought to do. But when they converge, the ethical challenge becomes really severe – perhaps even

---

35 Gardiner, 2011a.
36 Junger, 1997.

unprecedented. Hence, a "perfect moral storm." Perfect in its nastiness, anyway.

Nadia: What are these challenges? Or "storms" as you call them?

The Global Storm

Hope: The first is the *global storm*. Think of climate change from a strictly spatial perspective. Climate change is a genuinely global problem. Wherever on the planet greenhouse emissions are generated, they quickly get mixed into the atmosphere and start affecting the climate. As a result, they can have impacts anywhere and everywhere, potentially for anyone. This makes the climate system a global commons of sorts, which invites pernicious incentives. Plus, as we discussed, we lack good institutions for dealing with such problems.

Nadia: This just sounds like what we just talked about. How is the perfect moral storm analysis different from the tragedy of the commons?

Hope: There are similarities. But one difference is that the global storm highlights something that the tragedy of the commons framing obscures: deep issues of injustice.[37]

Eliza: What do you mean?

Hope: Well, one issue is that the tragedy of the commons assumes shared, you might say "symmetrical," vulnerability. In Hardin's example, notice that the presumption is that, if each herdsman keeps adding animals, ultimately they will all suffer the same harm: each one will lose their livelihoods when the pasture collapses. But in the global storm around climate, this isn't true. Instead, there are *skewed vulnerabilities*: those who are most vulnerable to negative climate impacts tend to be

---

37 Gardiner, 2011a, pp. 31–32; 118–120.

	the least responsible for causing them, and those who are most responsible tend to be much less vulnerable.
Nadia:	Can you give me an example?
Hope:	Well, Stacy illustrated this wonderfully earlier.
Stacy:	I did?
Hope:	Yes! You told us how many high-emitting, well-off Americans aren't yet experiencing the climate effects like Inuit communities, who emit little. This is true around the world. By and large, those most vulnerable to climate change are developing countries and marginalized communities. These populations have generally burned very little fossil fuel. By contrast, the rich countries – and the dominant groups therein – are overwhelmingly responsible for most past emissions, but much less vulnerable to the impacts, at least for a while. Moreover – and this is important – the poorer countries, being less powerful, are generally not in a good position to hold the rich to account. So, skewed vulnerabilities.
Nadia:	Can someone explain *why* these skewed vulnerabilities exist? Is it chance?
Stacy:	Part of it is just bad geographical luck. Climate change will likely cause environmental disruption most quickly in the Arctic, along the coasts, on low-lying islands, and in the tropics.[38] It just so happens that the less developed nations, poorer populations, and marginalized – often indigenous – communities tend to be concentrated in those places.
Hope:	But geographical luck *isn't* the biggest factor. In general, less developed, marginalized, and impoverished communities are more vulnerable *because* they are marginalized and impoverished. Being rich and powerful tends to mean having more resources to throw at preventing negative climate impacts. It also tends to mean having

---

38 E.g., see Bathiany, *et al.*, 2018.

better infrastructure to cope with impacts that can't be prevented: better health care, better insurance, higher-quality housing, and so on.

Andrew: That's exactly why climate change is sometimes referred to as an *amplifier*: it amplifies pre-existing vulnerabilities and inequalities.[39]

There is a momentary lull in the conversation as people digest this point.

Nadia: I'm struck by the fact that skewed vulnerabilities seem such an essential part of the problem, yet the tragedy of the commons analysis says nothing about them!

Eliza: I see what you mean. In the story, when the pasture collapses, all the herdsmen suffer equally: they have no place to graze their cattle. Similarly, people say that all countries and all people suffer from dangerous climate change. That makes it sound like what is happening is that countries are shooting themselves in the foot by not mitigating climate change.

Hope: Yep, but when you recognize the problem of skewed vulnerabilities, you see that wealthy, high-emitting countries – by exacerbating dangerous climate change – become responsible for climate harms they will not fully experience. Other countries and communities will pay a higher price. That's not a case of shooting oneself in the foot: it's more like shooting somebody else in the foot.

Nadia: That is quite different, isn't it? It seems seriously unfair and unjust.

Hope: I also think that some of the injustice goes pretty deep, perhaps with radical implications. For example, suppose that Bangladesh loses its coastal cities, and the Maldives loses its whole island territory.[40] This will happen mostly

39 Huynen, Martens, & Akin, 2013. It is also sometimes referred to as a threat *multiplier* (e.g., Werrell & Femia 2015).
40 These are both very real possibilities. See Kulp & Strauss, 2019.

because of the past behavior of those who refuse to take adequate climate action. In such a scenario, citizens of Bangladesh and low-lying islands like the Maldives *will* suffer, some will die. Even those who live long enough to relocate will lose the part of their identities tied to their geographical home. So, Bangladesh and the Maldives might be owed redress, such as compensation, from the high emitters. They may even be owed territory: a new home.

Andrew: That kind of loss makes my stomach churn. It shows how misleading it is to focus on the idea that, by not reducing emissions, the United States is shooting itself in the foot; it is more like we are shooting others in the head or heart. I agree that justice is an essential feature of the climate problem, and it's pretty bad that the tragedy of commons analysis ignores it.

A momentary silence.

Hope: Adama, you've been quiet so far. What do you think of all this?

Adama: I've been enjoying learning more about this topic, and especially how things look from your perspective.

Eliza: (*surprised*) Our perspective? Hmm. That makes it sound like we're missing some things. Are we? We all have blindspots, I'm sure. What are ours? What's *your* perspective?

Adama: Well, my perspective would take a while! I'd be glad to share it later. For now, let me point out a couple of things that jumped out from your discussion. One is that while I agree with Hope about skewed vulnerabilities, and with Andrew about climate change being an amplifier, I'd push those points further, or perhaps just be clearer about what they imply.

Eliza: Please, go on.

Adama: I think we can't truly understand climate justice unless we take seriously the background situation of wider injustice.[41] The existing world system is deeply unjust already, for a whole host of reasons, everything from the legacy of colonialism,[42] systemic racism,[43] the out-sized political influence of multinational corporations,[44] the bloody history of fossil fuel extraction ... you know what I'm talking about.[45] The point is that many climate injustices *build on* these and other background injustices.

Andrew: Absolutely!

Nadia: Can you give us an example? Something that might make the problem more concrete for the presentation?

Adama: Sure. Take "vulnerability." A lot of that is due to deeper forces. For instance, the fact that communities are where they are is often not "bad geographical luck," as Stacy put it earlier, but the result of a deeper history of injustice. Andrew already mentioned the vulnerability of indigenous peoples in Alaska, where 85% of their communities are coastal. But most of these communities didn't choose to be coastal; they were originally nomadic or semi-nomadic. They were moved to the coast by the Bureau of Indian Affairs, which wanted their settlements to be on navigable waters, so that they could barge freight and other stuff in.[46] In short, a central part of their current vulnerability was in effect imposed on them from outside, by colonial forces. You might even argue that if they were still nomadic, they would be much less vulnerable to climate shifts.

Eliza: Wow! I didn't know that.

41 Gardiner, 2011a, p. 119; Shue, 2014; Whyte, 2016; Whyte, 2017; Blomfield, 2019; Whyte, 2019.

42 Gandhi, 2019.

43 Quarcoo, 2020.

44 Kim & Milner, 2021.

45 Wenar, 2017.

46 These remarks are directly drawn from comments by an Indigenous Alaskan, quoted by Carr & Preston, 2017, pp. 765–766.

Adama: I'm not surprised. Most of us don't know much about how the way the world is now is a product of a messy *and not so distant* past. It's not something that our culture and education system has prioritized. Probably because much of it is pretty ugly. Nevertheless, it is crucial to recognize that many of the so-called "inequalities" out there are really in place because of background injustices.

Hope: Do you mean that there are real risks of *compound* injustices, where the powerful take further advantage of those already exploited under the current structure?[47]

Adama: In part. An obvious issue is international power politics. Think about the fact that, according to analysts, the Paris Agreement mostly benefited the United States and China, whereas Africa got the worst out of it.[48] That's bizarre, given the need to protect the most vulnerable, but it is also utterly unsurprising given the way the global system currently works.

Nadia: That's really depressing.

Adama: True; but we need to face it. We must also recognize that race plays a big role. Indeed, some say the skewed vulnerabilities you mention result in *climate apartheid*.[49] I've always found this framing striking. To me, climate adaptation is coming to resemble a highly unequal, segregated system, like South Africa's old regime. While wealthy, mostly white countries are most responsible for the climate problem, poor, largely black and brown communities will bear the brunt of the impacts, at least for a while.

---

47 Shue, 1996; Pogge, 2002.
48 Dimitrov, *et al.*, 2019.
49 Tutu, 2019. Apartheid was South Africa's segregationist legal system that discriminated heavily against non-white Africans and lasted from 1948 until the 1990s.

Nadia: That's really shocking. Can you give us another, more concrete example of how this plays out in practice? The one about Alaskan coastal communities was so helpful.

Adama: Sure. How about this? I see background injustice infecting even what issues get talked about. For example, so far today the discussion has been all about mitigation – emissions reductions and so on. We haven't talked about dealing with actual climate impacts – you know, questions like whose responsibility is it to fund adaptation efforts, and to compensate for real loss and damage on the ground. Unsurprisingly, ignoring this stuff is really common in developed countries, especially among the elites, since it seems likely that they bear a huge responsibility. By contrast, dealing with actual impacts is a central and pressing concern for most developing countries around the globe, and especially poorer and indigenous communities. It seems really unjust not to take that concern seriously. Yet it is routinely neglected in negotiations. I think that this is climate apartheid in action. If it were rich, white communities that were struggling to adapt, I'll bet that would be central to the international agenda.

Stacy: But surely without mitigation we're all doomed, so that is important to everyone. Shouldn't it be central no matter who you are?

Adama: You misunderstand me. I agree that mitigation is really important. But I'm saying that it is not the only thing that matters. Many poor and indigenous communities will be confronting severe, even existential, threats very soon, if they are not already. So, mitigation alone will not save them. They need action on adaptation, and on actual loss and damage – and they need it now. My point is that talking exclusively about mitigation reflects a bias towards the concerns of the rich, and the more developed nations. They are most worried about future catastrophe; but for many vulnerable communities catastrophe is not that far off, and for some it may be imminent.[50]

Stacy: We should be sure to bring up these issues in the presentation. Anything else you think we're missing?

Adama: Well, since you ask (*smiles*), the need to discuss impacts also affects how we talk about the science.

Stacy: (*interested*) How so?

Adama: Notice that even the idea of "dangerous anthropogenic interference" assumes that there's some level of interference that is not dangerous, presumably because people can adapt relatively easily. But then we have to be careful about whose ideas get to define what's dangerous.

Nadia: I'm not sure I follow you.

Adama: Sorry. Going a bit too fast, as usual! How about an example? Some of you have probably learned about the big dispute at the climate meeting in Copenhagen in 2009? Initially, the rich countries pushed for a 2 degree target for avoiding dangerous anthropogenic interference, right? But they were met with stiff resistance from a bunch of more vulnerable nations who thought that 2 degrees was a disaster for them. One delegate called it "a suicide pact for Africa" and some of the small island states pointed out that they'd be under water by then, due to rising sea level. So, these vulnerable nations wanted 1.5° degrees instead. Anything more was really dangerous *for them*, they said.

Stacy: I do remember that. Didn't the vulnerable nations win out ultimately? Paris does call for holding the global temperature increase to well below 2°C and pursuing efforts to limit its increase to 1.5°C, right?

Adama: True. That is at least a partial victory. But notice that the language is a bit of a fudge. They kind of agree that 2 degrees is too high, but then only commit to *trying*

---

50 Clare Heyward emphasizes the importance of adaptation, and the centrality of cultural loss. Heyward 2014, see annotated bibliography.

for 1.5. And the promise to try may not be that serious: some countries have said that they regard 1.5 as "aspirational" and 2 degrees as the more relevant target.[51]

Nadia: Interesting. I had wondered about the fuzzy language when Stacy first told me about Paris.

Adama: Still, my more basic point is that to understand targets, you need to understand impacts and vulnerability. But to understand those, you also need to think about justice.

Nadia: (*animated*) Let me just say that, as a nonexpert, I find this stuff really interesting and important. I've never really understood why climate change is a matter of environmental justice until now, and I doubt others you'll be talking to have either. I think you should emphasize it in the presentation.

Everyone nods in eager agreement, and Hope feels a swell of pride. They were taking ethics more seriously than she had feared.

Hope: I am so happy to hear you say that, Nadia. I believe we've moved way beyond the tragedy of the commons analysis already. I think we all see that skewed vulnerabilities and background injustice run through the climate problem understood at a global level. Still, sadly, the global storm is only one of our challenges. There are more storms to come; and once you recognize all of them, I think the tragedy of the commons analysis appears even more inadequate.

Nadia: Okay, what are these other storms?

The Intergenerational Storm

Hope: Well, if you recall, in the global storm we assumed that each state agrees that dangerous climate change is bad

---

51 For instance, during the Paris meeting John Kerry, then U.S. Secretary of State, reportedly described the 1.5°C target as "an add-on to the agreement, and more of an aspiration than a must-do," saying specifically "I think you can write that aspiration into the agreement in a way that doesn't make it the target or guidepost for the agreement" (Goldenberg, 2015).

	and is motivated to avoid it, typically on grounds of national self-interest?
Nadia:	Sure.
Hope:	What if that wasn't true?
Stacy:	Of course, it's true! Surely no country would *deny* that future hurricanes, forest fires, and droughts are bad. Right?
Hope:	Fair point. Any minimally moral person would probably grant that.
Stacy:	There you go, then.
Hope:	Not so fast. It's one thing to recognize something to be bad in the abstract, another to think that it's worth preventing.
Stacy:	I'm not sure I follow.
Hope:	We have to remember that climate change plays out over a long time period. What we do now will have effects for decades, centuries, even millennia. Given that, we might expect that fully responding to the problem would require states to be effective intergenerational stewards. They would have to be good at caring about, promoting, and protecting the interests of future citizens as well as citizens today.
Stacy:	(*hesitant*) And you think that's not true?
Hope:	I think it's a very bold assumption, if nothing else. Think about it. People alive in the coming centuries aren't even around yet. They don't have political power. They can't vote. Even many of the people who will exist later in this century are either young now, or have not yet been born. So, why think that current governments all around the world are good at protecting or promoting their interests?
Stacy:	When you put it like that, I suppose I agree. I'm not sure they would be.
Hope:	What worries me is that an alternative seems more likely. Let's take a pretty raw, simplified version. Suppose that,

at best, existing states are good at promoting the concerns of the current generation in power – say, people over 50. It's not obvious that this older generation will be sufficiently motivated to protect the future against dangerous climate change. If they are not, I don't think the interests of future people will receive much attention.

Nadia: What do you mean?

Hope: Well, suppose members of the older generation are largely interested in things that will happen while they are still around. In that case, they might act to prevent some really serious effects that are likely to occur in the next 20–50 years, but not much beyond that. So, for example, they might prefer to keep the various goodies of a fossil fuel economy for a few decades, even if that spells disaster later in the century and beyond. By then, they expect to be dead and gone. They might even be tempted to take risks of big impacts toward the end of their lives, so long as they can keep enjoying their goodies until then. We can't just *assume* that they'd want the same robust climate policies that younger people might want and need.

Eliza: So, you're saying that the older generation has disproportionate power, and they may choose to pass the buck to the future?[52]

Nadia: That's a horrible thought.

Hope: Agreed. The implications might be even worse. Notice that the problem doesn't arise only within one generation. Over time, each generation has to decide whether to take serious climate action. So, the intergenerational dynamic might explain why our grandparents' generation failed to act decisively in the mid-1980s, why our parents' generation is procrastinating now, and also why our generation might not do the right thing when

52 Gardiner, 2011a, ch. 5.

we take the reins of power. There's an incentive for each generation to reap the short-term benefits from a fossil-fuel economy while passing on many of the resulting costs to its successors. That might happen even if the shorter-term benefits are modest and the longer-term costs severe.

Eliza: So, you're saying there's a risk that successive generations, even as they suffer from the wrongdoing of their predecessors, might act in a similar fashion towards the future?

Nadia: Wow. Won't that make the impacts in the future get worse over time? I mean, because later generations suffer the consequences of the bad behavior of more and more generations before them.

Hope: That's the gist of it. The problem is called the tyranny of the contemporary.[53] I hope you see that it is different from the traditional tragedy of the commons. For instance, once again, the central dynamic is one of harming others, so that issues of justice loom large.

There's a pause. Hope shuffles nervously as she watches the others exchange looks of surprise, skepticism, and revulsion.

Andrew: I'm sorry, Hope, but the tyranny of the contemporary sounds way too dark. Doesn't it presuppose that human beings are monsters? What about love? If I didn't trust in love, I would have given up my climate activism long ago. It might sound like a cliché, but I'm sure that I'm going to love my children and want to protect them. I'm pretty sure that Mom and Dad, and Grandpa and Grandma love me. I think other people love their families too. Many also love their communities, their countries, and even humanity as a whole. We're not ruthless in the way your "tyranny of the contemporary" makes us sound. That's not who we are.

53 Gardiner, 2011a, ch. 5.

Hope: I believe in love! I also *agree* with you that most of us have strong intergenerational concerns, both for our own families and for the future of humanity itself. Still, some aren't so optimistic. They point to the massive failure of climate action since 1990 and say that some kind of intergenerational ruthlessness is the best explanation.

Andrew: (*winces*) Well, I hope that's wrong. But now I'm confused. If you believe people care about the future, why do you think there is a tyranny of the contemporary?

Hope: (*slowly*) My hope is that the main explanation for the tyranny we see is not generational ruthlessness, but something else.

Andrew: Like what?

Hope: Well, for one thing, I worry that the concern most of us have for future people can be easily overwhelmed by more immediate and visceral concerns, and so forgotten. More importantly, I believe there are few channels for turning intergenerational concerns into action, and certainly none that are effective on the scale we need. Think about it: elections are almost always focused on the next three to five years, and markets run on fulfilling people's immediate desires. So, in practice, other forces probably wash away most of the intergenerational concerns that many of us share. We need to find ways to fix that, to make concern for the future matter in the real world. In my view, that's the central intergenerational challenge: we need to fill the *institutional gap*.[54]

Andrew: I suppose that makes sense (*screws up his face in thought*). I guess I still don't buy this tyranny of the contemporary thing, though. Maybe our parents passed the buck to us, but the buck stops here. And not because we're better people, but because we're already seeing some big impacts! You know, the wildfires, for instance.

---

54 Gardiner, 2019a, pp. 206–207.

	And it's only going to get worse. Some keep saying this is the new normal. But *of course* it isn't. At the rate we're going, things are going to get much, much worse. Soon, these will look like the good old days. So, those around now should be motivated to act. There's no incentive for us to delay.
Hope:	Good points.
Andrew:	But doesn't that mean that the tyranny of the contemporary analysis has had its day? It might have been good for the first few decades of the climate problem, but it is basically out of date now, right?
Hope:	I can see why that view seems tempting. But I think it is more complicated than that. Crucially, it overlooks the fact that what happens in a tyranny of the contemporary can *evolve* over time.[55]
Andrew:	(*raising an eyebrow*) What do you mean?
Hope:	Well, intergenerational buck-passing doesn't need to take the form of *complete* inaction. It can also take the form of inappropriate action, like weak action or even strong action of the wrong kind. Even though some harmful impacts are upon us, and others are coming soon, we might still react to that in ways that are bad for the future.
Andrew:	Go on.
Hope:	Well, it is an empirical question ultimately, but I suspect that many of the impacts coming soon, over the next few decades, are probably already *in the cards.*
Andrew:	You mean that we're already committed to them at this point?
Hope:	Yes. Some warming is already physically locked in.[56] So, some future climate impacts will be too. For instance, I imagine that wildfires in the West and in Australia will continue over at least the next ten or twenty years. As

55 Gardiner, 2011a, chapter 6; Gardiner, in press.
56 Hansen, *et al.*, 2005.

Stacy said, it's almost certain that the global temperature will keep rising until at least the mid-century, and so there may be no way of mitigating our way out of whatever climate impacts that entails.[57]

Andrew: Hmm.

Hope: More importantly, further bad impacts might be *in effect* locked in for social reasons – you know, political, economic, even ethical reasons. Realistically, we might already be committed to those impacts because it turns out that we're not going to wean ourselves off fossil fuels quickly enough to avoid them. Fossil fuels are deeply embedded in our societies. So, it seems likely that we'll emit significantly more greenhouse gas before we're done, perhaps *much, much* more. Remember that even the more ambitious proposals being considered by countries right now are to get to net zero emissions only by 2050 or 2060.

Andrew: I think I see where you're going with this. If the older generation is focused on those impacts that are coming soon, and if many of those are in the cards, mitigation isn't going to help them. So, they might act on climate, but their priorities might be elsewhere, say on fighting wildfires and so on. Or at least that is the concern that is likely to be operative. It's the one that current institutions, being geared toward the shorter term, are best set up to take seriously, and probably at the expense of the further future.

Nadia: Wow. So, you're saying that even adaptation policy might start to manifest some kind of intergenerational buck-passing. That's really scary.

Hope: I'm sorry to say it, but it is.

Andrew: Heck, if it gets really bad, some generations might even be *justified* in having those priorities: saving people now

---

57 IPCC, 2021b, p. 17.

and letting the future fend for itself. It might be hard to tell people to cut down emissions for the sake of the future when their homes are on fire and their local economies are being devastated by drought or hurricanes. They might even say that they have a right to self-defense.[58]

Hope: That is a really worrying scenario. You make me worry that we might kick off something like *an intergenerational arms race*, where each generation tries to protect itself against harms in the short term – or in its own time – but in doing so makes matters worse in the future. That's a nightmare scenario we should really try to avoid.[59]

Eliza: Sobering stuff.

The Ecological Storm

Nadia: (*wincing*) Are there any more of these moral storms?

Hope: Two more.

Nadia: (*aghast*) Two?

Hope: (*apologetically*) Yeah. The next is the ecological storm. Climate impacts are spread across species, and there is more buck-passing there. Just as the costs of climate change are externalized to the poor in the global storm and to the future in the intergenerational storm, they are pushed onto animals and the rest of nature in the ecological storm.

Nadia: How does that work?

Hope: Well, we might worry about a *tyranny of humanity*.

Nadia: A what?

Hope: Do you remember that horrible old playground rhyme? You know: "the farmer kicks his wife, his wife kicks the child, the child kicks the dog."

58 For discussion of self-defense and climate change, see Chapter 2.
59 Gardiner, 2011a, ch. 6.

Nadia: I think I might have grown up in a nicer neighborhood than you! But why is the rhyme relevant?

Hope: With climate, the worry is that the current rich "kick" the poor, both "kick" future generations, and everyone passes the "kicking" on to other species, through the ecological systems on which they depend. In other words, the initial bad behavior sets off a chain reaction. At the end of the chain are not just the most vulnerable humans, but many animals, plants, and places. Think of the polar bears, the big cats, the rainforests, the Arctic. Moreover, if and when the natural world kicks back, it is likely to induce further cycles of buck-passing, creating a downward spiral.

Andrew: That looks plausible to me. We're already seeing tremendous costs of human expansion on the nonhuman world. Some suggest we're facing a sixth mass extinction.[60] Populations of wild animals have already crashed by 60%.[61] The Great Barrier Reef is at risk of total collapse due to coral bleaching, triggered by warmer water temperature. Meanwhile, species like the walrus, the pika, and innumerable others are facing extreme pressures as their ecosystems absorb a rate of ecological change not seen for millions of years. There's little doubt this will all get a lot worse if there isn't aggressive climate action.

Nadia: That's horrible. Animals are not at all responsible for climate change, and yet they seem like some of the most vulnerable to it. Talk about skewed vulnerabilities. Ecologically skewed vulnerabilities seem particularly bad.

Hope: Putting it like that makes me think about another ethical dimension of the climate problem. Just as high-emitting countries may owe vulnerable human communities

60 Kolbert, 2014.
61 WWF, 2018.

redress, perhaps humanity owes *nonhuman* communities redress.[62] Climate instability promises widespread suffering and death as species struggle to adapt. Humans did that. Do we have a moral obligation to reduce this suffering where we can? Ought we to help vulnerable wild species adapt to the new environment we have forced upon them? Again, it is not all about us shooting ourselves in the foot. The ecological storm involves a tyranny of humanity over the rest of the natural world as well.

Stacy: I understand where you are coming from, but does it make sense to say that we "owe" nonhuman species? Many animals lack consciousness or the ability to feel pain, more still rationality. I understand how ecosystem collapse can have very negative effects on humans, like Inuit communities. But I'm not sure the interests of animals matter much for their own sake.

Hope: These are profound moral questions. Do you think it is morally wrong to torture a housecat? Catch and torment a squirrel for fun? How about cut down the last redwood, or pave the rainforest?

Stacy: Such things make me uncomfortable, but I can't really say why.

Hope: I'm guessing that you feel uncomfortable because most of us believe that humans are not the only things in the world that matter for their own sake. Speaking for myself, if we cause great suffering for nonhumans, wipe out the polar bear, and destroy ecosystems, I can't help but think we are guilty of a profound ethical failure. Yet many conventional analyses of the climate problem completely ignore the value of nature. In my view, we will not meet the moral challenge of climate change unless we seriously consider how human behavior has impacted the nonhuman world.

Andrew: I am inclined to agree, Hope. I refuse to eat meat and dairy because I think the factory farming industry is morally abhorrent, but also because I would rather not use sentient animals as a resource if I can avoid it.[63] Still, I have to admit that I can't wrap my head around what it would look like to factor in the interests of all sentient creatures, let alone the entire nonhuman community, in a climate policy.

The Theoretical Storm and the Threat of Moral Corruption

Hope: To be honest, Andrew, neither can I. In fact, this is a great example of the final moral storm. It would be one thing if we face the other storms – global, intergenerational, ecological – with strong theory and robust institutions to guide us. But climate change involves the intersection of many areas where we have neither. We lack a good compass, if you like, and are more likely to be blown off course. This is the *theoretical storm*.

Eliza: What areas do you have in mind?

Hope: Well, we've seen that climate change brings together domains like intergenerational ethics, international justice, and the human relationship to nature. It raises questions about how to deal with scientific uncertainty, persons whose existence and preferences are contingent on the choices we make, how to value population and extinction, how to understand the value of nonhuman life, and so on. I think we're just not very good at all that stuff yet, and we certainly haven't worked out how to navigate it in terms of institutions and norms. For example, I have no really firm ideas on how to do right

---

63 For a philosophical dialogue on ethical vegetarianism in the Routledge series in which the present volume is a part, see Huemer, 2019.

	by the nonhuman world, even though I think we should try. And this not knowing is part of what makes climate change such a profound ethical challenge.
Eliza:	Fair enough. In a weird way, I'm glad to hear that you're struggling with those questions in philosophy too. In economics, we certainly are.[64]
Nadia:	Gosh, that's kind of disheartening.
Hope:	I understand why you say that, and I regret to say that the last part of the perfect moral storm might make you feel even worse.
Nadia:	What's that?
Hope:	Being in the perfect moral storm makes us vulnerable to moral corruption.
Nadia:	Moral corruption? Sounds insidious.
Hope:	In a way. What I mean by "moral corruption" is that we, the current generation, and especially the affluent in more powerful countries, are susceptible to ways of thinking and talking that are pretty distorted.
Nadia:	You mean fake news, alternative facts, and that kind of stuff?
Hope:	Partly. I think we're vulnerable to lots of tricks that undermine a good understanding of our problem. So, distraction, complacency, unreasonable doubt, selective attention, delusion, pandering, false witness, hypocrisy ....
Andrew:	Sounds familiar in the climate debate!
Stacy:	Yes! Unreasonable doubt has been a persistent problem for us climate scientists. It almost seems that the skepticism gets worse the more evidence we gather and the more confident we get in our conclusions. Surely it should be the other way around! We've also had to deal with all kinds of pressure – even abuse sometimes – that simply doesn't occur in most other areas of science.

---

64 E.g. Sagoff, 1988; Ackerman & Heinzerling, 2004; Jackson, 2011.

Hope: That is sad. Still, I have a deeper worry too. I'm concerned about our being tempted to favor ways of framing the climate problem that obscure many of its most important features. Given the storms we've discussed, we might expect ways of thinking and talking about it to emerge that gloss over, omit, or diminish important issues of justice and ethics. For instance, we, the relatively affluent, are at risk of taking perspectives on the climate issue that are warped by our own narrow concerns and don't take seriously the risks we impose on poor people, the future, and nonhuman nature. After all, those "stakeholders" – if you will – are usually not in a position to ensure that their concerns are registered and make a difference.

Nadia: Wait! That sounds a bit over-the-top to me. Can you give me some possible examples?

Hope: Sure. We've already seen some.

Adama: Yes. The tragedy of the commons analysis is really popular, even though it ignores skewed vulnerabilities, compound injustices, and the general historical context.

Andrew: Also, here's a dubious and quite convenient belief: that our governments reliably protect the interests of their citizens hundreds of years into the future. Any analysis that simply assumes that will invariably demand us to change less than I think we need to. It encourages complacency and so serves as convenient cover for more buck-passing.

Hope: Here's another morally corrupt idea: that climate change might not be a problem because humanity could always decide to live in huge domes, isolated from the rest of the planet.

Eliza: That does seem pretty fanciful! Especially if the idea is that we have the technology to do it for the whole of humanity in just a few decades!

Andrew: Notice also that it seems to assume that what happens outside the domes – to nonhuman animals and the rest of nature – doesn't matter, or at least isn't our responsibility.

Nadia: Okay. I now see what you all mean. Those framings really draw your attention away from important aspects of the climate problem. Interesting.

Hope: Let me raise an even more provocative idea. I think any suggestion that climate change is *merely* a practical, or a political, or a scientific problem risks moral corruption. Climate change is all of those things, of course, but it's also a profoundly ethical problem. For instance, we must consider *all* of the stakeholders, not just this or that country, the rich, the poor, the contemporary, or the human. As we said earlier, getting it wrong on climate doesn't only involve shooting ourselves in the foot. How did you put it, Andrew? Ah, I remember: it also risks shooting others in the head and heart.

Nadia: Well, maybe that analogy is a bit much. How about this instead? The traditional tragedy of the commons analysis invites us to think of ourselves as putting our own boats in peril, but when we turn to the perfect moral storm we see that we're also – and mainly – sinking other people's boats, now and in the future. The energy and shape of most of these storms comes from us; and they are powered by injustice.[65]

Stacy: (*taking a breath*) I'll admit this is a lot to process. I can't help but think you're correct to highlight the ethics stuff, Hope. I'll be thinking more about it, and how to

---

65 There's a complicated history of ship metaphors in environmental ethics. Garrett Hardin has a highly controversial piece on global population arguing that countries are effectively lifeboats, and that rich countries risk being swamped by immigration from the overflowing, sinking lifeboats of poorer countries (Hardin, 1974). Kyle Whyte offers a rival analogy for the case of climate change that opposes Hardin's metaphor, and artfully illustrates many dimensions of climate injustice (Whyte, 2019). See annotated bibliography.

integrate more philosophical thinking about justice into
my work as a scientist.

Hope: (*grins*) Wonderful! That's the goal!

Andrew: Yes. We *definitely* need to make clear in our presentation to the university that justice is right at the center of
the climate problem.

Stacy: (*laughs*) We got a little off our agenda, but it seems we
arrived at a productive take-away. I'm with Hope: climate change is an ethical problem, as well as a political,
economic, and scientific problem. Agreed?

The others *nod in unison. Hope beams. Sometimes thinking about
the difficulty of the climate problem had got her down, but not
today. She had convinced them. Maybe she could make a difference.
Maybe there was a way out of the perfect moral storm after all.*

Dialogue 2   Skepticisms

Late 2019

Part 1: Scientific Skepticism

Hope raps on the door of her childhood home. The door opens, and the face of her father appears. His smile, evident from the crinkle around his eyes, is hidden behind a greying beard.

Dad: Hope! I'm so happy you made it. (*winks*)

Hope: (*rolls eyes*) I'm only a little late.

Hope steps inside and is immediately engulfed by a heat that could rival a tropical island. She wastes no time tossing her backpack and dumping her coat on top.

Dad: Make yourself at home (*eying the pile, which Hope suddenly realizes appears quite out of place in her father's well-ordered entryway*).

Hope: Oh, stop. I'll collect my things when I leave.

Dad: (*chuckling*) I'm only teasing. I'm always happy to see you. Let me put the food in the oven. Please take a seat!

Hope obliges, sitting where she had for every meal during her childhood. Her father returns, wiping his hands on a towel.

Hope: I am sorry I'm late. I got caught up preparing for a meeting tonight.

DOI: 10.4324/9781003123408-2

Dad: (*with a tell-tale air of skepticism*) That climate activist stuff? (*Hope nods*) You seem to be spending all your time on it these days.

Hope: Pretty much! It's keeping me busy, but it's worth it. I've convinced the others that my work matters, which was a wonderful start. Then, our first presentation was really well received. Other student organizations have been getting involved, even the university administration.

Dad: Hmm. The science part must be difficult.

Hope: (*bracing herself*) Not really. Grasping the severity of the problem only takes a pretty basic understanding of the science.

Dad: (*scoffs*) If you say so. To be honest, Hope, I'm more skeptical than the greenie types you run around with. Even the so-called experts disagree.

Hope: (*annoyed*) The experts *don't* disagree ...

Dad: (*interrupting*) Anyway, I'm not sure why you're spending so much time on it.

Hope: (*Exasperated*) Dad! I spend so much time on it because climate change is a huge threat, to my generation, to humans everywhere, to the natural world, and in the near term especially to the poor and marginalized. It's so important! Can't you see that privileged people like you and me must use our advantages to try to fix the problem?

Dad: (*losing it a little*) C'mon, Hope. Don't you see that climate change is mainly a political bandwagon, eagerly jumped on by socialists and wacko environmentalists to further their agendas? You and your fancy university friends are caught up in a liberal bubble. You're out of touch with real people. Just listen to all those woke buzzwords! "Marginalized people." "Privilege."

Hope stays silent, stunned.

Dad: (*with a tone of regret*) Did I go too far?

Hope: Yes, actually. It's about time we had a serious conver-
           sation about this. Can we do that? Respectfully. That
           means no sarcasm.

Dad: I did go too far.

Hope: (*resentfully*) You did. You know I've been investing a
           lot of time – a bunch of my life energy – into learning
           about this. Can't you see *I* think it's very important,
           even if you don't? And – really – are you comfortable
           simply assuming that your own daughter is a wacko or
           in league with some conspiracy? Don't I deserve a little
           more credit?

Dad: (*slightly shaken*) I guess you're right; perhaps I shouldn't
           be so dismissive ... But we do have a disagreement here.
           I'm not going to pretend that we don't.

Hope: I'm not asking you to pretend. I'm only asking that we act
           like serious people, dealing with a serious disagreement.

Dad: (*defensively*)You have to admit that your lot can be
           preachy and self-righteous.

Hope: (*sharply*) And some of yours aren't unbelievably arro-
           gant and offensive? (*pauses, collecting herself*) Look, I
           dare say that there are difficult people and bad behavior
           on each side of the argument. That's pretty normal in
           political life these days. But that's not what we should
           be aiming at, right? You've always taught me to think
           things through and follow the arguments. Not to get dis-
           tracted or manipulated by cheap rhetorical tricks.

Dad: True. Honestly, I worry that you're being misled, and
           your good intentions exploited for political gain.

Hope: I'm afraid you are, too.

Dad: (*trying to lighten the mood*) Really? You think I'm being
           played by the special interests – the right-wing think
           tanks, funded by Big Oil, shady oligarchs, and their cro-
           nies? An unwitting dupe of global capitalism and the
           elites?

Hope: And you're worried I've been brainwashed by a bunch
           of loony lefties, using bad science to hijack the economy

and impose their hippie, Mother Earth, values on everyone else?

They both smile awkwardly, and the tension dissipates slightly.

Dad: So, how to begin?

Is Climate Science a Partisan Issue?

Hope: How about this. I'm guessing we both agree that science can become politicized and distorted in public debate, right? I remember you telling me that story about Einstein. You know, how he was always being interrogated by taxi drivers on the basics of relativity theory?

Dad: Right. The Nazi regime created a whole controversy about his work to discredit him because he was Jewish.[1]

Hope: It sticks in my mind because I'm always getting climate denier talking-points thrown at me in taxis! Interesting folks, those drivers; but some sure listen to a lot of talk-radio propaganda.

Dad: Why do you call it propaganda?

Hope: Propaganda aims to manipulate people. It presents information in biased and distorted ways in order to promote some cause or point of view.[2] Often it does this

---

1 Hamilton, 2013, pp. 96–99.

2 Here, Hope's conception of propaganda roughly corresponds to "information, especially of a biased or misleading nature, used to promote a political cause or point of view" (Lexico). The philosopher Jason Stanley has defined propaganda as part of a "mechanism by which people become deceived about how best to realize their goals, and hence deceived from seeing what is in their own best interests" (Stanley, 2016, p. 11). As a means of preventing the spread of propaganda, the Federal Communications Commission (FCC), the independent agency of the United States government that regulates public communication, established the fairness doctrine in 1949 that required television and radio broadcasters to discuss matters of public interest in a fair and balanced manner. Throughout its existence, critics accused the fairness doctrine of violating the constitutionally protected right to free speech. The fairness doctrine received many legal challenges throughout its life, and the FCC finally scrapped the rule entirely in 2011 (Matthews, 2011).

by being highly selective about which facts or values it mentions, and failing to take seriously any consideration that doesn't support its agenda. Some people are really good at propaganda. They can manipulate, entertain, and make you feel part of something bigger all at the same time.

Dad: Pah. Your side puts out propaganda all the time, ignoring scientific dissent and relying on your favorite experts.

Hope: I don't think so. But can we at least agree that manipulation is a problem, whoever does it?

Dad: Sure. I hate propaganda. It undermines proper debate and is a threat worth taking seriously in any free society.

Hope: Good; we agree there. So, let's turn to climate science. Do you accept that there are some basic matters of fact at stake there? I'm thinking of questions like: Is the global temperature rising? Are humans causing this? Is the problem going to get a lot worse? ... Do you agree that the answers to such questions are independent facts about the world, discoverable through science?

Dad: Of course.

Hope: So, these are not inherently political questions, right? I mean the real answer to a question like "Is the global temperature going up?" is independent of the politics of the day, and should be treated as such. For instance, it doesn't make sense to deny climate science *just because* you're on the political right, or to accept climate science *just because* you're on the left. It would be weird to think that there are two different *kinds of science* here – "left-wing climate science" and "right-wing climate science" – where people get to pick which science to believe based on their favorite tribe. It's not *inherently* a partisan issue. We've just made it one.

Dad: Agreed.

Hope: Good. So, if you reject basic climate science and I don't, what we disagree about is some facts about the world.

Dad: Yes.

Hope: Can we then agree about the best methods for resolving that disagreement?

Dad: Sure. We need scientific proof. You think we have it, and I think we don't. I think there's way too much uncertainty.

Who Should Judge the Science?

Hope: Good. But let's take this slowly. I'd like to start with the thought that actually you and I aren't the ones judging the science here.

Dad: What do you mean?

Hope: Well, neither of us actually does any climate science, or any kind of science at all for that matter. Neither of us has extensive training in it, or does cutting-edge research, right? Instead, each of us relies on testimony: what other people or groups are telling us.

Dad: Are you saying we're being manipulated? You by the scientists, and me by the skeptics?

Hope: Not so fast. So far, I'm just saying that we're relying on others. That's actually normal in science and everywhere else. When you and I say we believe in the theory of gravity, or plate tectonics, or the existence of other galaxies, we're relying on the scientific community for their views. We're trusting them, right? When we believe that Columbus came to the United States in 1492, or that the Romans invaded Britain, or J.K. Rowling wrote Harry Potter, we're also trusting others. Right?

Dad: I've never really thought about it like that, but I see what you're getting at. Neither of us ever actually saw J.K. Rowling writing *The Deathly Hallows*. We rely on others – indeed, the widescale testimony of others – in believing that.

Hope: Exactly.

Dad: But there are things we know more directly. Remember when we took that trip to Hadrian's Wall? We've seen evidence of Romans in Britain.

Hope: Well, we saw a wall and it looked old. We also saw lots of exhibits claiming to show Roman artifacts. But I guess for all I actually saw directly, the whole thing could have been faked, an elaborate conspiracy to trick us into believing that the Romans were in Britain.

Dad: (*shocked*) Why on Earth would you think that?

Hope smiles.

Dad: Oh, I see. Clever. You're saying that we believe an awful lot based on testimony.

Hope: I'm also pointing out that because of this reliance on testimony it is usually *possible* to be skeptical – like if you decide to think there's a huge Roman Britain conspiracy. Of course, usually those kinds of conspiracy theories are going to seem pretty out-there; you'd need strong reasons to get sucked into them.

Dad: Sure. In principle, I agree with that. But why think that my climate skepticism is akin to Hadrian's Wall skepticism? That doesn't seem very fair.

Hope: (*smiles again*) Why not? Suppose we are presented with a consensus account from a particular community. The historians, in the case of Roman Britain, or the scientists in the case of climate change. Normally, we – as nonexperts – would choose to accept that, right? We'd at least give them the benefit of the doubt. We wouldn't immediately jump in to side with rival groups – especially ones with vested interests, an ideological agenda, and no scientific credentials – particularly if nothing else is at stake.

Dad: Fair enough. But with climate I think there's *no* consensus and a huge amount *is* at stake.

The Scientific Consensus on Climate

Hope: Let's dig into it. Here are a few basic points about my understanding of the climate consensus. I'm listing them to see if we're on the same page. First, there's a huge amount that is just not controversial at all. No scientist disputes that there is more CO_2 in the air than there was fifty years ago or that the temperature is rising.

Dad: Really? But how do they know that?

Hope: Well, there's an observatory in Mona Loa, Hawaii, that has monitored the atmospheric concentration of CO_2 since 1958. When this record began, the concentration was 310 parts per million, but it has been steadily rising. Now it is 410.[3] The graph even has a catchy name, the Keeling Curve.[4] We also have indirect ways of measuring CO_2 content over much longer. Records from ice cores indicate that it is now higher than at any point in two million years.[5]

Dad: Hmm ... I hadn't heard that.

Hope: We also know that the global temperature is rising. We've been directly observing that for more than a century. For example, we can see that 2016 was the hottest year since 1880,[6] and almost all of the hottest years on record have occurred since 1998.[7] Other indicators show that the current decade has been hotter than all but one in the last 125,000 years.[8]

Dad: Okay, fine. But how do we know the increased CO_2 has occurred mainly *because* of burning fossil fuels? It could just be a natural cycle.

3   National Aeronautics and Space Administration, 2020.
4   Named after David Keeling, the scientist who did the initial research.
5   IPCCb, 2021, p. 9.
6   National Oceanic and Atmospheric Administration, 2021.
7   National Oceanic and Atmospheric Administration, 2021.
8   IPCC, 2021b, p. 9.

Hope: Scientists say that human activity – especially the burn-
           ing of fossil fuels – does a good job accounting for the
           data. In addition, they have strong reasons to believe
           it is not something else because nobody has found an
           alternative, natural explanation. They've looked, exten-
           sively; but none work.[9]

Dad: Let's say humans *have* increased the CO_2 concentra-
           tion. How can we be sure that causes the warming we've
           seen? That warming *could* be unrelated.

Hope: Dad, the greenhouse effect – the idea that, other things
           being equal, increasing greenhouse gases like CO_2
           increases temperature – is basic atmospheric science.[10]
           It's *much* less scientifically debatable than, for example,
           evolution. Remember – you always make fun of peo-
           ple who want evolution removed from school science
           classes.

Dad: But what if other things aren't equal? What if something
           counterbalances the increasing CO_2? There might be
           negative feedback loops: the increased greenhouse gases
           might trigger other processes that ultimately counterbal-
           ance the warming effect. I heard on a podcast that if the
           climate begins to warm then water will evaporate over
           the ocean and become clouds. The clouds would then
           reflect solar energy back to space, cooling the planet.[11]
           See, equilibrium!

Hope: Dad, that's not the whole story. It's complicated. Water
           vapor is also a *greenhouse* gas, so the vapor that doesn't
           condense into rain will have a *warming* effect. So, global
           warming will create more water vapor, which will cause

---

9   Mann & Kump, 2015.
10   Mann & Kump, 2015, p. 23.
11   This is called a *negative feedback loop*, where global warming has an effect –
like the creation of clouds – that, all things being equal, would cool the planet
(Mann & Kump, 2015, p. 25).

more global warming.[12] There are *positive feedbacks* too.

Dad: Hmm. So, we need to look at the *overall* balance, of both positive and negative feedbacks.

Hope: Yes. It is not a good argument merely to invoke *one* potential feedback mechanism in isolation. You have to look at the whole picture.

Dad: So, what's the state of the science there?

Hope: Scientists have now spent three decades looking into feedbacks and quantifying them. Nothing that's come up makes a big enough difference to offset what we're doing. In fact, as far as we can tell right now, overall the positive feedbacks outweigh the negative.[13]

Dad: Fine; I didn't realize that. But, *still,* leftists assume that global warming will have horrible effects. Why think that? *Even if* the release of greenhouse gases causes over-all global warming, how can we be sure that this change will be bad?

Hope: Not "leftists," Dad, *scientists.* Most of whom, I'm guessing from the meetings I go to, are hardly the radical type .... Do you know many atmospheric chemists and physicists? More like bank managers than bomb throwers, if you ask me.

Dad laughs.

Hope: Anyway, do you really want me to go through all of the negative effects? More drought, more extreme weather, melting glaciers that leads to sea-level rise and flooding, disease.

---

12 Mann & Kump, 2015. This is called a *positive feedback loop*, where global warming has an effect – like the creation of diffuse water vapor – that, all things being equal, would warm the planet *further*.
13 IPCC, 2018, pp. 4–6.

Dad: (*interrupting Hope*) Aha! But this is what I'm talking about. I heard that a recent study showed that the glaciers weren't actually melting!

Hope: (*sighs*) I think I know what study you are talking about. It was celebrated by climate deniers everywhere.

Dad: I'm not a *denier,* I just practice healthy skepticism. You should appreciate that as a philosopher.

Hope: I agree that *healthy* skepticism is a virtue. But there's a difference between that and denial. Proper scientific skeptics are trying to further science by raising important questions and pushing the discussion forward. They are responsive to evidence. Deniers are not.

Dad: You talk as if I am arguing in bad faith. I'm *trying* to listen to you, Hope.

Hope: Fair enough. I concede that if climate denial means bad faith, then you're not a denier.[14] Nevertheless, I still say denial because I think your skepticism is very selective. You cherry-pick. You ignore the overall – heavy – weight of evidence in favor of climate science. Instead, you

---

14   There are different senses of denial. First, the term "denier" might be read neutrally, as meaning simply someone who asserts that a given set of claims is false. However, in the climate context, richer meanings are often in play. Second, one might interpret "denial" in terms of what psychologists sometimes call "being in denial" – namely, "the refusal to accept reality ... [or] acting as if a painful event, thought or feeling did not exist," Denial is often taken to be a form of motivated reasoning bias, which enables one to reach the conclusion one already wants to believe (Miceli & Castelfranchi, 1998, p. 140). This is one of the most primitive human defense mechanisms (Grohol, 2016). It is in this sense that denial is not appropriately responsive to the evidence. There is a general consensus that motivated reasoning contributes to climate change denial (Kahan & Carpenter, 2017; van der Linden, et al., 2017, Kovaka, 2019, pp. 2361–2363). Moreover, importantly, motivated reasoning is very common across group divides and need not be a *conscious* form of self-deception (Kovaka, 2019, p. 2362). Third, some use the term "climate denial" in a stronger sense still to mean something like "the deliberate and deceptive misrepresentation of the scientific realities of climate change" (Dunlap, 2013; McKinnon, 2016, see annotated bibliography). While a useful definition for some purposes, Hope is not assuming here that denial need be deliberate.

choose to focus on the really minimal possible counter-evidence, and act like it is the most important thing. That seems more like simply refusing to accept reality than trying to improve the science through healthy skepticism.

Dad: Why do you think I'm cherry-picking?

Hope: Take the study you just mentioned! It only claimed that *East* Antarctica specifically *might* not be losing mass overall.[15] Also, the lead author made *absolutely clear* that his findings did not contradict the consensus finding that, yes, Antarctica as a whole is losing mass in a warming world. A healthy skeptic would never conclude that this single study poses a deep challenge to the scientific consensus.

Dad: Maybe you're right about that study. But what's wrong with airing doubts? Shouldn't we be able to speak our minds?

Hope: In general, but there are limits. It's not okay to shout "Fire!" in a crowded theater, for instance. In the climate case, I don't think people should be shouting "there's no fire!" when there's ample evidence that the flames are spreading and will soon be out of control.[16] In particular, I think there are very strong ethical duties *not to mislead* people about evidence of serious potential harm, especially when the risk is of severe, even catastrophic, harm.[17]

Dad: Maybe. But I'm not trying to mislead; I just think there's less certainty than you are suggesting. Are you 100% sure that climate change is happening and that it will bring catastrophe? C'mon, Hope. Even scientists acknowledge that there's uncertainty.

---

15 Hall, 2017.
16 McKinnon, 2016, pp. 2122–14.
17 Brown, 2012, p. 104.

Hope: *Of course there's uncertainty.* That's why scientists offer climate projections that cover a full range of scientifically realistic scenarios. This full range is what the IPCC summarizes, and which serves as the basis for the scientific consensus.[18] Your problem is that you are so overrun by skepticism that you refuse to acknowledge the evidence stacked against you; for example, the 97% of scientists that accept climate change is happening.[19]

Dad: Of course you'd bring up the 97%! But I've heard that was greatly exaggerated; it's closer to 80% consensus.

Hope: First of all, the 97% number has been checked multiple times and substantiated.[20] But what you just said reveals the real issue: why on earth do you think that *80% agreement* among scientists *isn't enough* to believe that burning fossil fuels threatens us with dangerous climate change?

Dad: The 20% might still prove the 80% wrong!

Hope: Maybe the 3% will .... But remember, there are big incentives within science to prove hypotheses wrong: you can make a career out of it. Yet so far nobody has disproved climate change, despite years of trying. That's pretty important given that overturning the consensus would probably result in Nobel Prizes, given how strong the current science is. In short, the fact that no climate scientist has been able to disprove the consensus tells us a lot.[21]

Dad: I admit that a lot of what you say sounds impressive. Some of it is news to me. But I've heard of lots of disputes about the science in the news on TV and online. I haven't raised even a *quarter* of their objections to the science. So of course, I'm skeptical.

18   E.g., IPCC, 2021b, p. 16.
19   Oreskes, 2004.
20   Cook, *et al.*, 2016.
21   Oreskes, 2019.

Hope: You need to be careful, Dad. The thing with climate deniers is that they are experts of the gish gallop.

Dad: What on earth is that?

Hope: It's a debating technique where you try to overwhelm an opponent with arguments without regard for their individual accuracy and strength. Often it is done by sprinkling in just enough tiny bits of truth to make it difficult for the opponent to rebut specific points without going into lots of details. The technique is apparently named after a famous creationist, Duane Gish. It is called a "gallop" because part of the strategy is to run through points really quickly, so that people don't have a chance to see that the overall position is basically a castle of sand. It can work pretty well on people who don't know much about a given topic. It can also throw those who do off balance for a while if done effectively.[22]

Dad: I don't think *all* the skeptical arguments are weak.

Hope: Many are *really* weak; they could be righted by a simple internet search. I can also see that some would initially seem strong *if* you didn't know that they had already been addressed in the scientific literature. One characteristic of this type of argument is that it is imperviousness to progress: a Gish galloper keeps raising issues that have already been successfully addressed, but refuses to acknowledge that.

Dad: I don't think you're being fair. Perhaps some of my arguments were less strong than I thought, but there are good arguments against climate change! Some of the "climate denial" – if you insist – is from folks from MIT and Princeton. How do you explain that?

The Evidential Standard of Absolute Certainty

Hope: Well, let me say again that I don't think we can reasonably expect either of us – you or me – to resolve the

22 Bokuluch, 2013.

scientific disputes, especially all by ourselves. Neither of us are scientists, so our task is different.

Dad: And what task is that?

Hope: One of epistemology and ethics.

Dad: Huh?

Hope: Epistemology is the philosophy of knowledge and justified belief. It involves figuring out what to believe, and how strongly to believe it. Ethics involves figuring out what to do and what kind of people to be.

Dad: What do you mean?

Hope: Well, our place in all this is not scientific, but social and political and at bottom ethical.

Dad: (*with a fake tone of surprise*) So! You're playing the ethics card?

Hope: Suppose for a moment, just for a moment, that mainstream science is right: humanity then faces an enormous threat, possibly even an existential threat. Billions of lives are at stake, and perhaps the future of humanity itself.

Dad: I agree that that would be a big deal. I also agree that it would put a big responsibility on those of us alive now, especially in rich countries to do something about it. Let's even say it's a moral question.

Hope: Good. Aren't we getting on well today? (*smiles*)

Dad: But you haven't made the case yet! I don't believe that the threat is real! What about all that disagreement? You seem to be evading me.

Hope: I'm not evading you. I've already set out the consensus. And I'm about to say more about the disagreement stuff.

Dad: Go on.

Hope: You said, "all the disagreement"? But I've just said that it's not all that big. The consensus seems to show that.

Dad: Okay, but say more about that, and about the prominent scientists who disagree.

Hope: Sure, but just one more thing. You and I agree that it would be a big deal if climate science were true or reliable. When the stakes are that high there's a question about what sort of proof we're looking for.

Dad: Hold on. What do you mean "what sort of proof"? I just mean that it should be proven. There's a lot of money at stake in decarbonizing our economy. I don't want us to be wrong and end up wasting all those resources tackling a fake threat that doesn't really exist. Scares have happened before. Think about when Leonard Nimoy raised that whole scare about global cooling! Or the Y2K bug. These were a bit before your time, but they were both completely overblown.

Hope: Actually, Dad, there's good reason to think that part of the reason the Y2K bug didn't cause chaos was *because* people took it seriously behind the scenes.[23] But, that aside, of course there will be instances where potential threats turned out to be overblown; however, we have to remember that it's also happened before that we've ignored early warnings and then paid a big price. Think of Neville Chamberlain's policy of "appeasement' in the 1930s, and how it contributed to the rise of fascism. And sometimes we've averted catastrophes by acting early, before all the science is in, like with the ozone hole. Scientists suspected chlorofluorocarbons (CFCs) were to blame for the ozone depletion, and were confident that this would ultimately result in increased rates of skin cancer and cataracts; but they didn't yet have overwhelming evidence that CFCs were responsible. Imagine if people were as skeptical about the role of CFCs as they are about climate change!

Dad: I guess sometimes alarmism is justified. But not always.

Hope: I wouldn't call it "alarmism." It sounds like you are trying to say there's something irrational about it. But

23 Uenuma, 2019.

you're agreeing with me that sometimes it is appropriate to recognize a severe threat and be motivated to respond to it, right?

Dad: Sure, sometimes.

Hope: So, the real question is whether or not we are justified in raising the alarm about climate change.

Dad: Point taken. But I don't think it is justified given the uncertainty.

Hope: I'm guessing what you'd really like is to be *absolutely certain* that climate change is happening and will get really bad if we don't act. 100%, as you said earlier.

Dad: Yes.

Hope: How often *do* you demand absolute certainty?

Dad: All the time, if it's important.

Hope: But that's not true, Dad! Remember how you said earlier you didn't deny evolution? Even though there is an overwhelming scientific consensus that evolution is real, there's not absolute certainty.

Dad: (*hesitates*) Well, there is an incredible amount of evidence for evolution, and I'm pretty sure the scientific consensus is greater than in the case of climate change. But you're right. I'm not absolutely certain.

Hope: Right! Neither am I. But I'm used to uncertainty. I'm a philosopher.

Dad: What do you mean?

Hope: You know the quote "I think, therefore I am" from Descartes?

Dad: I've heard it, yes. I didn't know it was Descartes. And I'm not sure I understand what it means.

Hope: Well, you can take my word for it. I'm a philosopher. I know about Descartes the way archeologists know about Roman walls and climate scientists know about climate change.

Dad: (*rolls eyes*) Sure, sure. Where did you get that tongue?

Hope: (*grinning*) Descartes tried to figure out if there's anything we know with absolute certainty. He tried to get there by what he called the method of doubt.[24] The idea was to set aside anything that he could possibly be mistaken about.

Dad: Sounds like a reasonable method.

Hope: It does at first. Unfortunately, Descartes quickly got into some trouble. Specifically, he concluded that the only thing he could know for sure is that he existed when he was thinking, or more accurately that thinking was occurring: "I think, therefore I am."[25]

Dad: Wait! Philosophers think that's *all* we know?

Hope: Typically, no. It seems a silly view of what we know. But my point here is that it is perhaps not such a silly view of what we know *with absolute certainty*.

Dad: What do you mean?

Hope: Descartes began by asking himself whether he could possibly doubt the evidence of his senses: that there's a table here, a plate there, and so on. He realized that he *could* doubt these things. He realized that he often dreamed things, like eating a meal with a lovely guest, and sometimes these dreams seemed very real. Sometimes in the dream, he even persuaded himself that he was awake, even though later he woke up and realized his mistake.

Dad: I get that. I'm pretty sure I'm really sitting with you at the dinner table, but I admit that I could be dreaming. (*a dawn of realization*) Dinner!

Dad jumps up and scrambles to the kitchen. Hope leans over for a peek and catches a glimpse through the doorway of him taking a pan out of the oven. After a minute or two, he returns.

Dad: Well, our dinner is a bit toasty, but we avoided total catastrophe.

Hope: Glad you acted in time!

---

24 Descartes, 2008 (first published in Latin in 1641), pp. 14–17. For further analysis of Descartes and uncertainty in climate science, see Gardiner, 2011a, pp. 457–463. Also see annotated bibliography.
25 Descartes, 2008, pp. 17–24.

If Hope's dad caught her intended double meaning, he didn't let on. They eat in silence for a few moments.

Dad: So, this Descartes fellow. Did he arrive at certainty?

Hope: Well, he eventually decided that he couldn't trust his senses. You know, because he might be dreaming.

Dad: Wait a minute. That's a big jump. It is one thing to say that you might be dreaming from time to time. It's quite another to think that you never have sensations of real things. Maybe you'd never even be able to dream of a dinner table if you never experienced one.

Hope: Very good. We'll make a philosopher of you yet.

Dad: (*guffaws*) God forbid.

Hope: Okay. You're right. So, Descartes also asked whether it was possible that he could be dreaming all the time. Say, if some evil genius were deceiving him. These days we might ask if it's possible that we are all plugged into a virtual reality machine. You know, like the Matrix or some other computer simulation.

Dad: That's really out-there!

Hope: I agree. But we're just asking whether it's possible.

Dad: This is really bizarre. I guess I have to concede that it's *possible* that I'm living in a computer simulation, even if it's very improbable. So, I see what you're getting at. Absolute certainty is a very high standard of evidence. In fact, it's the wrong standard. If I insist on absolute certainty, I'm going to end up like Descartes struggling to prove much to myself beyond that I exist.

Hope: Exactly.

Dad: So, your wider point is that science can never be certain, and so climate science can't either.

Hope: And that's not because there's anything wrong with climate science! It would be equally true if I demanded absolute certainty about evolution or the existence of the Roman Empire.

What Standard of Evidence *Should* We Adopt?

Hope: (*continuing*) But my more specific point is that you should not insist on standards for climate science that you wouldn't insist on for other, similar things. If you're making a special case out of climate science, then there's something very fishy going on already, before we start talking about actual scientific arguments.

Dad: So, what's the alternative? What do we normally do?

Hope: Well, we accept different standards of evidence for different purposes. I've been learning that in law there are a variety of standards: beyond reasonable doubt, clear and convincing evidence, preponderance of the evidence, probable cause, reasonable suspicion. Beyond reasonable doubt is a very high standard. Preponderance of the evidence is less so. Probable cause is even more lax.

Dad: Yes! In criminal cases, it's innocent until proven guilty. It would be terrible to throw an innocent person in jail. So, beyond reasonable doubt is the right standard.

Hope: Exactly. We insist on a higher standard because we think that there's more at stake if we're wrong. This is where the ethics comes in. The threshold of evidence ought to depend on the cost of being wrong.[26] So, the standard is lower for making an arrest or getting a search warrant, since the burden on the accused is assumed to be lower. Of course, in theory the evidence should still be substantial. The police must demonstrate probable cause.

Dad: So, which is the right standard for climate science?

Hope: Good question. Specifically, what standards of evidence must we meet before climate action is justified?

---

26 In philosophy of science, this risk of being wrong is known as *inductive risk*. The argument from inductive risk challenges the value-free ideal of science by pointing to the moral costs associated with wrongly accepting or rejecting a hypothesis (Hempel, 1960, see annotated bibliography; Douglas, 2000). For a thorough discussion of values in science, and implications for policy, see Douglas, 2009.

Dad: I'm going to be demanding.

Hope: I'm guessing that you are. But I'm not sure in which direction.

Dad: What do you mean?

Hope: Do you remember when I was 13 and wanted to date that guy in high school?

Dad: The one I didn't know? Who was two years older than you?

Hope: Right. You asked me all about him. You also talked to the parents who knew him, including those of girls he'd taken out before.

Dad: I did. Wouldn't any decent parent do the same?

Hope: I certainly understand why you did.

Dad: Good.

Hope: But suppose most of the parents had advised against it. They said they'd had negative experiences of him, or their daughters had. But some said that they didn't see a problem and thought that he was probably okay. Would that have satisfied you? Would you have allowed me to date him then?

Dad: No way! I wouldn't have let you go if even one parent had warned me off!

Hope: So, a preponderance of the evidence, or even reasonable suspicion against him, would have been a deal breaker for you?

Dad: Yes. I'm not going to take that kind of risk with your life. That boy needed to pass a "beyond reasonable doubt" test that he wasn't any kind of threat to you. Honestly, I'd prefer "beyond unreasonable doubt"!

Hope: But, don't you see? You think you had a duty to protect me from bad outcomes, so you demanded strong evidence that he wasn't dangerous before I could see that boy.

Dad: I think I see where this is going.

Hope: I'm *guessing* you think you have a similar duty to protect me, your grandchildren, and great-grandchildren against serious, possibly catastrophic threats. So, how much evidence do you need before you take the climate threat seriously? You said you wouldn't need that much to want to protect me from my high school crush!

Dad: (*sounding uncertain*) I'm not sure.

Hope: Why not? I've told you about the scientific consensus, and you yourself said that its probably at least 80%. Why isn't that more than enough for you? If 80% of parents had warned you off that boy, you never would have let me date him.

Dad: Let's step back for a second. You're saying that the evidence for climate change passes the thresholds of reasonable suspicion and preponderance of the evidence.

Hope: I actually think it passes "beyond reasonable doubt" as well. But let's not get distracted. My main point is that you are insisting on some really high standards – probably way beyond "beyond reasonable doubt" – *before* being willing to do *anything* to protect against the climate change threat. I think that's really irresponsible. Even reasonable suspicion and preponderance of the evidence ought to be enough to get you moving.

Dad: ... like it would on protecting you from dating a guy with a bad history.

Hope: Indeed. You also have another reason to be suspicious. It's plausible to think that what's actually going on in the climate case is that the deniers and their friends in the fossil fuel industry are exaggerating the minimal level of remaining disagreement so they can keep making money in the short term.[27] And you, and many others like you, are buying it.

Dad: You mean like what the cigarette companies did?

27 Oreskes & Conway, 2010. See annotated bibliography.

Hope: Yes! In fact, some of those fanning the flames of climate denial are the very *same people* who insisted that cigarettes did not cause lung cancer. They are funded by vested interests now just as they were then.[28] Of course, we know better now about cigarettes. But many people died too young before the manufacturers admitted the danger involved in smoking.

Dad: That does sound dodgy. But climate skepticism is different.

Hope: Let's try another example. Suppose the mainstream military establishment thinks Russia, China, or al Qaeda, whoever, poses a threat to the United States. How much disagreement would you need before you think there's no *threat* or at least that it's not worth our while to protect ourselves against that threat?

Dad: A lot!

Hope: A few political hacks with an axe to grind wouldn't do it?

Dad: Definitely not.

Hope: Nor would a bunch of PR from folks funded by pro-Russian trade associations?

Dad: (*laughs*) Are you kidding me?

Hope: So, you get how unreasonable it is to trust a potentially biased source, then! In the climate case, you should be cautious about believing whatever the fossil fuel companies say, right? Especially when they fund the same people who used to manufacture tobacco denial?

Dad: But what about the *scientists* who disagree?

Hope: There aren't many. You mentioned a couple at MIT and Princeton who are skeptical, but there are plenty of scientists at those places and all around the world who disagree with them. Also, even the few scientists who are

28 Oreskes & Conway, 2010.

skeptical about some part of the science accept most of the consensus.

Dad: So you say; but some don't! Anyway, the scientific majority isn't always right. It's not a popularity contest. As we agreed, there are real facts of the matter, independent of what anyone thinks is the case.

Hope: Sure. But in this case what we've got is a *tiny* minority of scientists, most of them not even climate scientists, denying that there is a climate problem. Moreover, most of them are not even denying that.

Dad: What do you mean?

Hope: At most, they're saying it's not quite as big a problem as the mainstream suggests. Or they're saying there might be cheaper ways of dealing with it, and so on.

Dad: What's your point?

Hope: Let's go back to the military analogy. Suppose you've got a very small number of people who are experts on international relations trying to tell you that the threat coming from China or Russia is not very serious. But almost all of them are not experts on China or Russia, they are experts on other things. And most aren't saying that there's *no* threat from China or Russia, they're just saying that maybe the threat is different than widely believed, or there's an easy way to deal with it.

Dad: All right.

Hope: In that case, I doubt you'd be much moved by the argument that we shouldn't do anything to prepare for the potential threat, right? In particular, you wouldn't argue that we should ignore the 97% of experts raising the alarm just because 3% disagree.

Dad: I see where you're coming from. I acknowledge there is no way I would give such credibility to such a small group of people in the case of national defense. I'd say we were foolish not to make some preparations and in fact I'd insist on pretty serious preparations. It's a price worth paying to be secure.

Hope nods.

Dad: So, your point is that, if I say that in the national security case, about China, Russia and other potential threats, then, I should say the same thing for climate?

Hope: Yes! I'd also point out that the connection between national security and climate change is not merely metaphorical. The Pentagon has said that climate change will increase global instability and conflict, as well as threaten defense infrastructure.[29] They consider climate change to be one of the top threats to American national security.[30]

Part 2: Skepticism about the Role of Ethics

Dad: Wow. I'll admit, you've given me a lot to think about. You may even have changed my mind about how seriously we should take the science. But hear me out, Hope. I still think the standard of evidence needs to be higher than you suggest. The cost of taking "climate action" and *being wrong* would be enormous. We would devastate our economy for nothing! That would be horrible not only for the fossil fuel industry or old fogies like me, but also the youth and future generations! I hope you don't think so badly of me that you believe I've never thought about what's best for you and your children.

Hope: Don't worry. I have faith that your heart is in the right place.

29 Department of Defense, 2010, pp. 84–88; Department of Defense, 2019. For example, in a recent op-ed for *The Hill,* Lieutenant General Norman Seip said, "assessing and addressing the threat of climate change is critical for the future viability of our force" (Seip, 2018).
30 Klare, 2020.

National Self-Interest

Dad: I'm worried that we'll end up having to sacrifice a lot of what we care about, and much of what makes America the place that it is. If what you're saying is right, then we need massive transformation. Many things will be lost. Fundamental freedoms may be sacrificed. I want to be very sure before we go in that direction. That's what Bush senior meant when he said, "the American way of life is not up for negotiation." It's also why Trump insists that he's the President of Pittsburgh, not Paris.

Hope: Let's talk about some of those issues. But, as a preliminary, I'm a little surprised by your doomsday attitude. You're usually such a technological optimist, with great belief in the American ability to innovate and adapt. But you seem to think that we won't be able to do that in weaning ourselves off fossil fuels. Do you think we're forever stuck with coal and oil? That the American way of life is inconceivable without them?

Dad looks thoughtful.

Hope: Why can't we transition to clean energy? Why would such a transition threaten our way of life or our fundamental freedoms? I don't think that America would stop being America just because we'd be driving *electric* cars and trucks, any more than it stopped being America when most of us stopped riding horses. We're way more resilient than that. You have a thing about the millennials being "snowflakes," but you are starting to sound pretty snowflaky to me!

Dad: (*rolls eyes*) I'm certainly no snowflake, but I guess I hadn't thought about it that way. Still, my problem isn't about change alone, it's that climate action will hurt the economy. How are you going to pay for climate action? By raising our taxes, no doubt.

Hope: Well, you seem to think taxes are fine for some purposes. You often support increasing military expenditure, even though it is already very high. Why is this different?

Dad: National defense is necessary for protecting our country.

Hope: So is combating climate change!

Dad: But what about our national *economic* self-interest? It looks like the Paris agreement will be very expensive for the United States. Isn't Trump right when he says that he's the president of Pittsburgh, not Paris? That he should put America first and withdraw from crushing climate treaties?

Hope: I don't honestly think that withdrawing does put America first. How is severe climate change in the interests of the United States? What if Trump had said he was "the President of Paradise, California"? That went up in flames in part because of climate change. Or what if he'd described himself as the President of Miami, which likely would end up underwater? I don't see how it is in the long-run interests of the United States not to be part of the solution.[31]

Dad: Fine. Maybe we should do something. But why join an international treaty, and subject ourselves to external requirements we might find unreasonable?

Hope: Addressing climate change requires global cooperation, so playing nice is needed. Plus, the United States played a leading role in putting together the Paris treaty, as it did most of the previous agreements. So, it's a bit ironic for us to act like these agreements have been imposed on us by outsiders, as part of some sort of conspiracy. We basically designed them, and their structure reflects our preferences.[32]

Dad: Okay; but I still think you're being unrealistic. Countries are simply not going to agree to anything that's not in

31 Gardiner, 2017c.
32 Dimitrov, *et al.*, 2019, p. 3.

their interests. For a treaty to work, it has to be in the national interest of all those who sign it.

Hope: What do you mean by "national interest" here?

Dad: I'm just saying that countries will do what they want, what fits their objectives. They always do.[33]

Hope: But isn't "doing what you want" different from "doing what's actually in your interests"? People often choose to do things that are very bad for them. Say I'd gone drinking the night before the LSAT. I doubt you'd have said that was in my interests, even if I'd wanted to do it.

Dad: True.

Hope: Things get worse when you start talking about countries. Countries do lots of things because their leaders decide to do them. But the mere fact that a leader chooses to do something does not automatically make it good for the country. If Kim Jong Un decides to declare war on the United States, that's not good for North Korea *just because he said so*. And the same is true for any country, including ours. A President might *say* they are acting in our interests. But they can be *wrong*, seriously wrong.

Dad: Fine. I suppose I meant that the United States will not accept a climate plan that's not in its short-term economic interests.[34] That's how countries usually act.

Hope: I'm not sure that countries always act in their short-term economic interests. Remember the Great Recession! We kind of shot ourselves in the foot with our out-of-control housing market. And what about the Brits with Brexit ...

Dad: I mean countries *should* do what is in their short-term economic interests.

Hope: Really? Even if it conflicts with their long-term interests, including their long-term economic interests?

Dad: (*looks uncertain*) At least in most cases.

---

33   Posner & Weisbach, 2010, ch. 3.
34   Weisbach 2016, p. 188; Posner & Weisbach, 2010, ch. 3.

Hope: You don't mean that. Maybe it would be in the *short-term* economic interests of the United States for us to sell all our weapons technology to China, Russia, and Iran. But I doubt you think that's really in our national interest in the long term! Sometimes your long-term interests can be undermined by trying to make a quick buck.

Dad: Hmm.

Hope: It also occurs to me you haven't really done anything to *show* that exiting Paris – or failing to act on climate more generally – is even in our short-term economic interest *overall*. You simply assert it. But there's at least a discussion to be had about the balance of costs and benefits, and what counts as a relevant economic interest. For instance, some studies suggest that, as well as costs, there are major short-term benefits to decarbonization: efficiency gains, health benefits, wider social benefits, and so on.[35]

Dad: Maybe. But suppose we change tack and focus on long-term economic interests. Shouldn't we look to the economists for guidance? Won't they tell us what is in our long-term interests? And they can do that without us having to go into the ethics and justice stuff you want to bug me about.

Hope: I can see that some economics might help. But I'm not sure you can exclude ethics.

Dad: Go on.

Putting a Price on Carbon

Hope: The consensus within economics is that we need to regulate carbon, probably by putting a price on it. Typically, economists agree that a big part of the climate problem is the way some people are externalizing their costs on

---

35 Low Carbon Monitor, 2016, pp. ix–xi; Grubler, *et al.*, 2018; for discussion, see Lenferna, 2018, pp. 221–223.

to others. "Passing the buck," as I like to say. Well, basic economic theory tells us that we need to change the incentives so people make better decisions. For example, we need to internalize the costs of polluting behavior by putting a price on carbon so that those who do it pay the costs. If we impose a carbon price that reflects the true social cost of climate change, then people wouldn't pollute as much. They would have an incentive to shift away from fossil fuels. The carbon price is an adjustment for market failure – the failure of the actual market to register the true costs of fossil fuels, which is what an ideal market would do. The theory is straightforward. It's basically ECON 101.

Dad: Hmm. What does putting a price on carbon look like in practice? I've heard that the costs of climate damages will be pretty low. Even if we should do something about climate change, it needn't be very much. Just a few dollars a ton for the CO_2 we use. I could maybe get behind that.

Hope: Unfortunately, the true cost of carbon is probably not that low. Your friends are cherry-picking the numbers.

Dad: Why do you say that?

Hope: Suppose we focus on what policy people call "the social cost of carbon": the cost of the damage done by emitting an additional metric ton of CO_2. On the one hand, almost all economists agree that the number is positive, and so we should put a price on carbon. A few years ago, the National Academy said the best number was probably about \$42/ton of CO_2.[36]

Dad: That's quite a bit more than I thought. And on the other hand?

Hope: The range of estimates is enormous. A recent literature review gave a mean cost of around \$55 per ton of

36 National Academy of Science, 2017, p. 2.

CO_2, but the range was from a cost of nearly $2400 to a *benefit* of about $13 per ton.[37] Another recent study, based on admittedly pessimistic assumptions, suggested that the damages caused by each ton of CO_2 emitted might be much higher – in the range of roughly $2,700 to $205,000, with a central estimate of around $27,000.[38]

Dad: What! Anywhere between being slightly beneficial to a cost of over $200,000 per ton? That's a huge range!

Hope: Yes. It is the sort of range that suggests that the real action may be more about what's behind the numbers than the numbers themselves.[39] Still, for now, maybe we should focus on the fact that there's a lot of agreement that it should *at least* be in the range of $40–$200 or so. I say this because imposing a climate price like that would shift incentives dramatically, in a way that would justify a lot of climate action.

Dad: You may be right. Still, I'm not even exactly sure what we're talking about. How is the social cost of carbon number used?

Hope: So, in our country the federal government is required to do a cost–benefit analysis on new projects and policies. For instance, if you have a new project like a new power plant or a dam, and you're trying to figure out whether it's worthwhile, you have to add up the costs and compare them to the benefits of going forward. If the benefits are greater than the costs, then, other things

37 Wang, *et al.*, 2019, p. 1506.
38 Archer, *et al.*, 2020. In the paper, the numbers are expressed in per ton of carbon, not carbon dioxide, so that the range is "$10,000 to $750,000 with a central value of about $100,000 per ton C" (Archer, *et al.*, 2020, p. 2083). Here, we converted to the carbon dioxide equivalent by dividing the carbon by the ratio of its molecular weight to that of carbon dioxide. This is 12 units C to 44 units CO_2 (12/44 or 1/3.67) (EPA, n.d.).
39 Fleurbaey, *et al.*, 2019, pp. 90–96.

being equal, you are justified in proceeding. If the costs are greater than the benefits, you are not.[40]

Dad: That sounds sensible. How does the climate cost fit into this?

Hope: Well, for example, if a particular regulation was projected to reduce CO_2 emissions by 1 million metric tons in 2020, the estimate of the value of its CO_2 emissions benefits in 2020 given a social cost of carbon of $42 per ton would be $42 million dollars.

Dad: Interesting. How has the U.S. government employed the social cost of carbon?

Hope: The Obama administration used the $42/ton number from the National Academies. By contrast, when the Trump administration was forced by the courts to have some number, it went for $1–6/ton.[41] I assume that was the minimum they thought they could get away with.

Dad: A low number sounds more reasonable to me. But let's get back to what drives these huge differences in estimates. What's at stake there?

Hope: Interestingly, things that will bring us back to ethics and justice.

Dad: You're kidding me.

Hope: Not at all.

Dad: I suppose I shouldn't be surprised (*wryly*). Go on.

Hope: Well, to construct a climate economic model you need to make a lot of basic assumptions. You need to assume that the global economy will keep growing more or less as usual through the climate changes that occur, or you need to predict it will not.

Dad: I doubt that there will be smooth economic growth if we hit the big climate shifts you are talking about. The next generation would probably be worse off than we are.

40 Frank, 2005. "Cost–benefit analysis" can be understood in different ways. For more discussion, see Chapter 6.
41 Nuccitelli, 2020.

Hope: That would be my guess too. So, you might be surprised to learn that the $42/ton number is based on the assumption of continued strong economic growth.[42]

Dad: Huh.

Hope: There are also assumptions about substitutability. Economists tend to assume that whatever we lose from climate damages can easily be replaced by something artificial. But I'm not sure. There may be some things we can't replace. Fresh drinking water, for example, or water for crops. In a severe drought, there may be nowhere feasible to get it from.

Dad: I suppose.

Hope: Doesn't assuming substitutability seem a little complacent when it comes to severe climate impacts? The whole economy – all of human life – relies to a great extent on the natural world. This is why you don't need to be a tree-hugger to care about the environment: "ecosystem services" supply huge economic value to humans, from water purification to building material.[43] In fact, scientists say that the value we get from nature is massive – probably much greater than the value of the whole of the human economy itself.[44]

Dad: (*grinning slightly*) So you *don't* think we could simply go live in giant domes?

Hope: (*missing the humor*) Maybe eventually. But our technology is very far from that at the moment. I doubt that we can make the switch – house 7 billion people and all our agriculture – in a few decades, or even a century. That's fanciful thinking.

Dad: Okay ...

Hope: Still, what actually makes the biggest difference to the economic models is our attitude to the future. Whatever

42 The National Academy of Sciences, 2017, pp. 70–74.
43 See Sagoff, 1988, especially ch. 5.
44 Constanza, *et al.*, 1997.

choice you make about this more-or-less determines your results: it basically overwhelms the rest of the model. Moreover, the choices basically depend on questions of ethics and justice.

Dad: Really? How's that?

Hope: Well, one example is how much consideration we think we owe future generations. Are we entitled to value our own interests much more highly than theirs just because they are *our* interests? Does it matter whether the future is richer or poorer than we are? Is each generation entitled to an equal share of resources, or is what matters how big the pie is, whoever gets it? How much do we value population growth? Do we care about increasing the risk of human extinction?[45]

Dad: Stop! You're making my head hurt.

Hope: I sympathize. Welcome to my world!

Three Generation Model

Dad: That all sounds like a bit of a mess to me. What if we stop focusing on trying to get precise numbers – maybe that's going to be misleading whatever numbers we pick. Instead, can't we just go with a much more crude judgment? Maybe what I want to stress is the interests of my generation, our children, and our grandchildren. That's what I'm getting at when I'm thinking about acting in the national interest.

Hope: Notice how we've gone through at least three different notions of self-interest at this point and they're really different. We've had "whatever the country actually wants or decides to do at a particular time," then "short-term, narrow economic interests," and now we have this talk of three generations: you know your generation, your

---

45 See chapter 6 for a longer discussion of economic discounting and intergenerational ethics.

children (*smiles widely, pointing at herself*), and then my children. Surely we can't just keep wandering from one to the other, picking whichever one suits you at a given time.

Dad: Fair enough. So, let's stick with the "three generation model."[46] That seems like it comprises our national self-interest, and what citizens and their representatives want.

Hope: On that point, I think you're a bit over-optimistic.

Dad: How?

Hope: Well, your original idea was that climate policy is constrained by national self-interest. But do you really think that governments *always* act in the interests of you, your children, and grandchildren? That administrations reliably promote the interests of their people over at least three generations? That just seems so implausible to me. It's almost quaint.

Dad: (*scowls*) All right, settle down. Maybe it is not what they always do, but I'd say they do it sometimes. For example, one justification for the protection of Yellowstone National Park was to conserve the land for future generations.[47] In any case, protecting three generations is what governments should do.

Hope: So, now you're admitting that your self-interest claim isn't some grand *empirical* claim about how the world actually works? Instead, it's a *normative* claim about how things *should* be, and an aspirational one at that?

---

46   Although beyond the perspective of Hope and her father's conversation, the reader should be aware of the "7th Generation Principle" of the Iroquois Nation. While the Three-Generation Model may be forward-looking in the context of contemporary politics, it is quite parochial compared to the ancient Iroquois philosophy that "In our every deliberation, we must consider the impact of our decisions on the next seven generations." See Seven Generations International Foundation.

47   National Park Service, n.d.

Dad: (*grumpily*) Fine. But I still think it is a view about self-interest, not ethics. It is self-interest in an expanded sense.

Net Zero Argument

Hope: How do you think this three-generational motivation would play out for climate change?

Dad: Suppose for a moment that we agree with your scientists.

Hope: They are not *my* scientists!

Dad: Sheesh. *Our* scientists ...

Hope: Thank you.

Dad: I think we can get all the climate action we need out of the three-generation concept of national self-interest without needing any ethics.

Hope: Hmm. You seem to be claiming at least two things. The first is that the self-interest of you, the children, and the grandchildren all *converge on the same policy*; and the second is that that policy is good for the *long term of humanity* as well.[48]

Dad: Isn't that obviously true? If you are right about how bad climate change might be, it's good for everyone if emissions go to zero as quickly as possible. The fact that we all care about three generations is enough to motivate us to get to zero quickly.

Hope: I'm not sure. The fact that each of us cares about our own children and grandchildren doesn't imply that we all care about the *whole* generations coming after us. And some people might not care that much even about their own descendants.

Dad: But how much would that matter?

Hope: It could matter quite a lot. Let's think about this carefully. What do you mean by getting to zero "as quickly

48 Gardiner & Lawson, 2021.

as possible"? We could do it tomorrow if we wanted to. We could just turn everything off.

Dad: Not that quickly. That would cause chaos and lots of harmful impacts. It is clearly bad for lots of us around right now, including me and you, and I don't think that's in our self-interest.

Hope: So, when? Next week? Next year? Next decade? Next century? When should we be shutting down all the power plants and dismantling the fossil fuel infrastructure?

Dad: Well, if you are right about the science, I'm not against starting now. Why don't we stop building new fossil fuel infrastructure? And replace the old stuff with greener infrastructure as it wears out?[49]

Hope: What about replacing infrastructure even *before* it wears out? Isn't the threat great enough to justify that?

Dad: I'm not sure. It's costly to build new infrastructure. I think countries won't shut down power plants that are already built until they've run through their life cycles. That's not in their interests.

Hope: How long does energy infrastructure last?

Dad: Fifty years easily, perhaps as long as a hundred.[50]

Hope: Wait! So, you're not saying we should move *all that quickly*. You're willing to keep all the existing infrastructure going until it wears out: maybe 50–100 years for some of it?

Dad: Sure. Any other view is just fantasy. Countries won't stop using their existing fossil fuel infrastructure until it wears out. It's not in their economic interest to retire it. So, we're stuck with those emissions. I'm just suggesting we should not build any new fossil fuel infrastructure.

Hope: Ah. Two points.

Dad: (*grimaces*) Just two? Go ahead.

---

49 Weisbach, 2016, p. 185–186. See Gardiner and Weisbach, 2016, annotated bibliography.
50 Weisbach, 2016, p. 187.

Hope: First, you seem to have gone back to the short-term eco-
          nomic account of national self-interest. So, we've got
          two conceptions of national self-interest in play again:
          the short-term economic one, and the three-generation
          model.

Dad: Hmm. Maybe I did sort of do that.

Hope: The second point is that there seems to be a conflict
          between them. Suppose you stick with the short-term eco-
          nomic version of the self-interest constraint, and assume
          (as you did just now) that this means that existing energy
          infrastructure has to run its course. Then, we're in really
          serious trouble. Notice that China alone has built a lot of
          power plants in the last ten years or so. It will take many
          decades for those to wear out. Presumably, that means it
          is going to take a lot of years to reach zero. So, I don't see
          how accepting the short-term economic constraint gets
          us – humanity – to where we need to go.

Dad: So, you see a potential conflict between the interests of
          the generations then?

Hope: Yes. It might be in your generation's interests to keep
          the fossil fuel plants going. But if the climate impacts are
          really severe, it's probably not in mine or your grand-
          children's. If we're going to protect all of us, we need
          more aggressive action.

Dad: Hmm. Maybe we need to take a wider view. How
          about this? Clearly, there's a way in which it is not
          in my *real* interests that you and the grandchildren
          suffer severe climate change, even if it is in my short-
          term economic interest. Maybe we need to reject the
          short-term conception and go deeper with the three-
          generation view.

Familial Self-Interest?

Hope: Say more about that. How do my interests and the inter-
          ests of my children get bound together with yours? How

is this version of the three-generation view supposed to work?

Dad: Do you even have to ask?

Hope: (*smiles*) Only for philosophical purposes.

Dad: You must know that you're a big part of what I care about, even who I am. You having a good life, pursuing your goals, chasing your dreams. These are all a part of what makes my life a success. If things go badly wrong for you, my life is much worse than it would otherwise be. I'll feel the same way about your kids when they come along, and a little bit like that about the success of our community, and America too. I want the country to go on, to prosper. If it doesn't, I see my own life as implicated as well, as an American. I count it as a kind of failure on my generation's part, and so in some way on mine too.

Hope: I don't think everyone feels the same way.

Dad: No; but a lot do.

Hope: Anyway, you're starting to sound like a certain kind of ethicist.

Dad: How so?

Hope: What you started out calling self-interest has now expanded to include lots of other people: family, friends, community, the country, perhaps even humanity. It is hardly a selfish doctrine.

Dad: Sure, but those are things I want – maybe if I didn't care or had fewer connections, it would look different.

Hope: Perhaps. But my point is that, if I understand you correctly, you are *not* saying that you want your family, friends, and the country to prosper *because* it benefits you when they do. You're not valuing them as a means to anything else, and in particular because they bring some extra goodies for you. Instead, you're saying that you want them to prosper because *you value them for their own sake*. You want them to flourish, as Aristotle would say.

Dad: That sounds right. I can see why that looks more like ethics to you, but I still see it as self-interest. Besides, talking about self-interest will get more traction with other people than ethics I think. That's why I'm skeptical of your focus on climate justice.

Hope: I think there's a lot of confusion around the term "self-interest." But would you agree that, at the very least, we seem to have come a long way from the original idea that there is some kind of feasibility constraint based on short-term, narrowly economic self-interest?

Climate Extortion

Dad: I still think there's something to that idea. What about this? Surely, the United States has less at stake in climate change than many other countries. We are richer and not so vulnerable.

Hope: So, you think it's okay for us not to act? I don't follow you.

Dad: No. We should act. But there's a fundamental constraint: a treaty needs to benefit us more than it costs us. We won't do more than is in our interests, so we'll need something extra. Other countries, those with more to lose, will need to induce us, give us an incentive.[51]

Hope: I'm not sure about that.

Dad: Which part?

Hope: All of it, actually!

Dad: You're going to have to be more specific.

Hope: For a start, I don't see that other nations have much more at stake than we do, at least in the long run. A shift in global temperature of 4.5 degrees or more in 50–100 years is basically Armageddon for us and everyone else. I don't see how the current American way of life survives that, with or without trucks. Even 2–3 degrees

---

51 Posner & Weisbach, 2010, p. 86.

looks pretty bad for us. We might see less suffering and death in the short term. Maybe our wealth and infrastructure helps there. But even that's a little misleading. People are losing their homes, their livelihoods, and even their lives already.

Dad: Suppose – for the sake of argument – I agree with all that. Even so, we have *less* to lose than other places. Some countries are a lot more vulnerable. You know, places like Bangladesh would lose a lot from sea-level rise; some small island states might literally disappear. To get enough climate action to protect the most vulnerable, other countries will have to help us out. Make it worth our while.

Hope: Again, I'd say we're also vulnerable. Think about Miami, New Orleans, Manhattan ...

Dad: Maybe. But I still think we're relatively less vulnerable overall, especially than places like Bangladesh, the Maldives ...

Hope: It sounds kind of callous to be so indifferent to their plight. But I have another worry. I think you might be encouraging us to act like extortionists. I'm not sure that many Americans will want that. I'm not sure that they will think of it as in their self-interest to become the climate mafia! It doesn't fit very well with your expanded, and morally loaded, concept of self-interest.

Dad: Wait! What on earth are you talking about?

Hope: Suppose for a moment you're right that the United States is less vulnerable, and countries like Bangladesh and Tuvalu are more so. Your argument was that we, Americans, should be compensated for reducing our emissions to protect them.

Dad: If you want a successful climate treaty that everyone can get behind, that's what needs to happen.

Hope: But you seem to be proposing that they – Bangladesh, Haiti, the Maldives, and others – get together to *pay us off* to stop emitting so much. You're making it sound like we should run a protection racket. We're threatening them with our high emissions, and saying we'll only stop if they "compensate" us. But that sounds like demanding money for menaces. Doesn't behaving like that make us the climate mafia?[52]

Dad: Hold on. What are you saying?

Hope: (*energetically*) In fact, it actually might be worse than that! In law, we often adopt a "polluters pay principle" for environmental damages. But you seem to be advocating a "victims pay principle." That's bad enough. But then notice that the victims benefit more from robust climate action *because* they are more vulnerable, and one of the biggest reasons they are more vulnerable is *because* they are so poor. In effect, your view ends up promoting the functional equivalent of a "vulnerable pay principle" and a "poorest pay principle."

Dad: Wait! I didn't say any of that! You're making me sound like a monster!

Hope: Sorry – occupational hazard! I wasn't trying to do that. I was just following through on the logic of what I thought was your background argument: that we should get inducements from the more vulnerable because we're less vulnerable.

Dad: I see the problem.

Hope: Anyhow, that pure self-interest *could* lead to such extortion seems to make clear the importance of *fairness*, and that justice may demand that the United States goes against its own "best interest."

Dad: I'm still not sure.

Hope: I have another objection. I'm not convinced your proposal is actually feasible. Suppose the poor countries tried to pay us off to stop emitting, do they even have

---

52 This section draws on Gardiner 2016b and 2017b.

enough resources to make it worth our while? And if they did have something we wanted – diamonds, uranium – wouldn't we probably find other ways to get it? Isn't there likely to be an easier, cheaper way than giving up our fossil fuels? Like invading them, or propping up some tinpot dictator who would hand it over for a bit of military support and a few million in a Swiss bank account?

Dad: You have a grim view of international relations.

Hope: Grim? You are the one preaching hard-headed realism and short-term self-interest.

Dad: Touché.

Hope: Anyway, my point is that your "feasibility" argument doesn't look that feasible. I don't see how you get to zero quickly, or who is supposed to pay to protect the most vulnerable if we don't and they can't.

Dad: Hmm.

Hope: Let me raise one more thing. All this talk of "feasibility" is a little misleading.

Dad: How?

Hope: Usually, what we mean by saying that something is infeasible is that it is literally impossible. Like when we say that it is physically infeasible for me to jump to the moon, or economically infeasible to give every American a billion dollars. But on your account the real reason that robust and fair climate action is "infeasible" is that countries like the United States *simply don't want to do it*. You seem to be saying that we, as Americans, *are our own feasibility constraint.*[53]

Dad: Ah.

Hope: This is part of what makes your approach seem like extortion. You have us demanding a pay-off – "compensation" – for not threatening others through our

---

53 G.A. Cohen raises a related but not identical point regarding those who claim a socialist society is "impossible" (Cohen, 2009, ch. 4).

behavior, something that we shouldn't be doing in the first place. Aren't we better than that? I think we are.

Self-Defense

Dad: How about this? I can see some demands we might make would be unreasonable, even extortionate. But some wouldn't. We can't be expected to sacrifice our interests *completely*, even to save poor people in other parts of the world. There must be some limits.

Hope: Something like a right to self-defense?[54] Are you thinking that we have a right to protect ourselves against harms caused by climate policy, as well as harms caused by climate change itself?

Dad: Something like that.

Hope: I wonder ... I'm guessing the conditions will be tight.

Dad: What do you mean?

Hope: Normally, we don't think you can do just anything to defend yourself. For instance, there needs to be proportionality. If I shoot my neighbor just because I'm afraid he's going to step on my flowerbed, I can't claim self-defense.

Dad: How is that relevant here?

Hope: Well, it seems disproportionate to invoke a right to self-defense to justify continuing with high-emission American lifestyles when some countries are being threatened with severe drought or going underwater.

Dad: Point taken. Geez, you're really not going to back down, huh? No matter what I say, you insist that the United States has a moral responsibility to take pretty extreme action on climate change, and soon.

Hope: Extreme action? To me, it's the threat that's extreme, especially to my generation, our children and grandchildren. Weaning ourselves from fossil fuels seems pretty

54 Gardiner, 2016b, pp. 122–125. See annotated bibliography.

tame by comparison. That's why I'm so committed to my activism. It's why other young people are doing everything they can to raise the alarm and hold the older generations accountable. Like Greta Thunberg; yet for her efforts, she gets labeled an extremist, and demonized by some.

Dad runs a hand through his beard, contemplating Hope.

Dad: What can I say? Much of what you've said is quite compelling, I admit. I just encourage you to be sensitive to the fact that not everybody is as persuaded by the science as you are, or thinks America has a special responsibility to clean up a global problem. Using the three generations model and national self-interest and incentives may get you further than tree-hugging, hippie rhetoric, and talk of climate justice. Nevertheless, you've convinced me that your work is important. You're good at this.

Hope: (*sighing*) Thanks, Dad. I guess. I'm not sure how you can hear everything I've been saying and still dismiss me as a hippie. But I suppose we've made progress.

Dad: (*laughs*) Hey, Rome wasn't built in a day – or should I say Paris! A few more family dinners and maybe you'll make me as radical as Greta!

Hope gives him a pained smile and takes a bite of lightly burned brussels sprouts. She certainly has her work cut out for her.

Dialogue 3   Individual Responsibility

2019, Later the same day

Part 1: Individual vs Collective Responsibility

Hope's frustration churns within her as she cuts across the university lawn on her bike. Sweat trickles down her forehead and onto her clenched lips. Her feet pump the peddles with furious energy, wind whistling in her ears. She can't stop thinking about her argument with her dad. Did he come around? To some extent, maybe. But not entirely, even after all that effort. And there are so many more like him that will need convincing if they are really going to hold back climate change. On top of it all, she is now late for a meeting of the climate group.

Standing, Hope loops one foot over the back tire and rides the last few dozen feet on one peddle. She breaks in front of the rack and hops off. Fingers made clumsy by her rush and irritation, another minute slips away while she secures the bike. Finally, she lets the lock fall with a clink and rushes into the building.

Eliza and Andrew are already there, chatting animatedly. They look over as she enters. She forces herself to return their smiles. If she can't even persuade her own father of the urgency of climate action, how can she convince the entire student body? Tonight, she needs to figure out the answer to that question.

Hope: (*with a grimace*) Sorry I'm late!
Eliza: No problem! We were just strategizing.

DOI: 10.4324/9781003123408-3

Andrew: Yeah, we were trying to figure out what the goal is for today.

Hope: Isn't the plan to make a list of things the school and the student body should be doing to fight climate change?

Andrew: Yeah. Eliza and I were just debating what the list should look like.

Hope: (*laughs, feeling her irritation dissolve*) Well, that's what we're here to figure out, right?

Eliza: Of course. I was just telling Andrew that I'm not sure we should include things about what students should be doing individually to reduce their carbon footprint.

Hope: (*surprised*) Why not?

Andrew shrugs, but Eliza seems not to notice.

Eliza: I think it's condescending, elitist, and honestly misses the point. My emissions don't make a difference, and climate change is largely a structural problem anyway. I also fear endless, counterproductive in-fighting. So, I suggest focusing on establishing goals for the university, not for individuals in their private lives.

Andrew: And I just don't get why you think that! Individuals need to do their part by reducing their carbon footprint and adopting climate-conscious behaviors. We should be making clear to the students that they have these responsibilities.

Eliza: I disagree. The responsibility falls on the school as a whole!

Hope: Woah! I haven't even sat down yet.

Eliza and Andrew grin sheepishly. Hope shrugs off her backpack and pulls up a chair.

Hope: Okay, let's work this out. Eliza, why do you think it's condescending to talk about what individuals should do to reduce their climate impact?

Is Critiquing Carbon Footprints Condescending?

Eliza: It feels so "holier than thou." Take an example. I once went to a demonstration against the fossil fuel industry.

I spent almost a full day making protest signs. Not just enough for myself, but also some extras in case anybody else needed one.

Hope: That was nice of you.

Eliza: I thought so too! Anyway, the protest was two hours away by bike, and so I drove. When I got out of the car, this other protestor immediately denounced me to his friends as a hypocrite, as part of the problem – just because I drove there! I was so upset I left the signs for others to take and went home.

Hope: I'm sorry. That does seem counterproductive. In fact, it sounds to me like he was probably just being a jerk. What my Dad would call "Virtue signaling" – you know, trying to impress people by acting all morally superior, but in a disingenuous way. I bet he cared more about looking good than actually doing good. I doubt his carbon footprint is anything to crow about.

Eliza: In a society like ours, it is difficult for anyone to have much to crow about footprint-wise.

Andrew: Still, what if someone really is making a big effort with their own emissions, and trying to set an example for others? Isn't it okay for us to strive to do better and to hold others accountable as well? Sending signals is important when you're trying to drive social change. Surely you admit that individuals should be trying to drive less, right?

Eliza: I'm not sure. Even if they should, I don't think we should be focused on criticizing others for their emissions. If you think reducing your carbon footprint is morally required, reduce your own.

Andrew: Do you think its condescending to tell other people not to litter? Or to stop smoking around non-smokers?

Eliza: (*hesitates*) I suppose not. But that's different.

Hope: How?

Eliza: I guess the biggest reason is that people emit carbon mostly as a byproduct of doing the usual, everyday things. You know, trying to live a normal life.[1]

Andrew: But we're trying to change what's normal, right? We need to!

Eliza: Sure. But my point is that, *given the current infrastructure*, many emissions are necessary if people are to live their lives. So, it seems unfair to criticize. By contrast, nobody needs to litter, or to smoke around others; so, moral criticism there is justified.

Hope: So, you're saying moral criticism is not necessarily condescending, but it is condescending when the criticism isn't fair?

Andrew: But why isn't it fair to criticize someone for living a carbon-intensive lifestyle in a climate crisis?

Eliza: Like I said, most people don't have much choice in the matter. They emit carbon – often, lots of it – just by living. They can't control the shape of their cities or what energy sources power their grid. They have to live where they are. For example, some students live far away from campus because they cannot afford to live closer and need to drive to get to campus.

Andrew: That's sometimes true, I suppose. But always?

The Question of Carbon Elitism

Eliza: Mostly! You can't get away from carbon emissions unless you pretty much go off the grid entirely. Building houses, harvesting food, heating your home, charging your devices … All of it produces greenhouse gases in our society. … (*hesitates*) And there's another thing.

Andrew: Yes?

Eliza: Honestly, I feel like most of those criticizing other people for their carbon footprints are well-off white

---

1 Sinnott-Armstrong, 2005, p. 301; Jamieson, 2014, pp. 155–158.

people – folks who have enough privilege to let themselves off the hook pretty easily, without much real sacrifice. They just buy themselves a new status symbol – say, an expensive hybrid or electric car – and call it good. That kind of thing. They don't realize – or don't care – that calling for the less well-off to avoid driving is very different; it is likely to impose serious burdens on them, and in ways that are unequal and unfair.

Hope: Hmm. I see your worry, Eliza. I also feel the pull of your point about setting things up so that some can "buy their way out." That does seem elitist.

Andrew: (*thinking aloud*) I hear you, too. I'm also guessing that most of us privileged folks do well in some areas, but badly in others. If you can afford the electric car, chances are you are flying off somewhere exotic in the summers too. Overall, you may do worse than someone with a cheap, efficient car who doesn't fly much.

Eliza: So, do you agree that we shouldn't try to generate a list of individual behaviors, since it encourages elitism?

Andrew: Not so fast! Most students at this school are well-off and white! How is it elitist to tell *them* to reduce their carbon footprints? Did you know that the richest 10% of the world population are responsible for about half of household consumption? That itself accounts for well over half of total climate pollution globally.[2] Some call this group the "carbon elite."[3] They possess significant personal or family wealth, and so are very likely to emit *a lot* in avoidable ways.

Eliza: Let's be a little careful here. I agree that *some* are rich enough that *some* reductions come easy for them. Maybe that's *your* experience. Still, the richest 10% *globally*? Isn't that individuals making more than $38,000 a

2 Nicholas, 2021.
3 Nicholas, 2021.

year?[4] In my hometown, that's not a flashy electric car and vacations in Tahiti. Not even close.

Andrew: Ouch! Point taken. (*puts hands up in mock surrender*) Nevertheless, my privileged experience may still be relevant. I suspect a good chunk of *our* student population have parents in the *top 1%* globally. That's anyone earning over $109,000 per year.[5] I know there are 38 colleges in America where *more* students come from households in the top 1% of the income scale *nationally* than the bottom 60%. That's more students from families earning over $630,000 annually than from those earning less than $65,000 annually.[6]

Eliza: Wow. I didn't know that.

Andrew: Amazing, isn't it? It raises a lot of questions. However, my point here is that it's safe to say that a large chunk of our student body can leverage their wealth to change their lifestyles. If they choose not to, the barrier seems to be cultural more than anything else. Extravagant, carbon-intensive lifestyles garner social status; people feel like they "need" expansive wardrobes and vacation homes just to keep up, and fit in. I can understand why the carbon elite – people like me— may be reluctant to give them up. But surely it can't be elitist to say we *should*?

Eliza: Point taken. But that doesn't change the fact that plenty of students here come from families nowhere near the top 1% nationally. Remember we're making a list of responsibilities *schoolwide*. So, we're supposed to be talking to everyone. I'm not sure it's helpful to blunder in telling students whose families are struggling to send them here that they are the "carbon elite." The issues are more nuanced than that.

Hope: Perhaps we should focus more on opening the conversation, rather than very specific directives? How about,

4 Capstick, *et al.*, 2020, p. 63. Calculated using purchasing power parity.
5 Capstick, *et al.* 2020, p. 63.
6 Aisch *et al.*, 2017.

instead of telling the students never to drive, we say, "when possible, take a bike or public transportation instead"?

Andrew: We should also emphasize that certain climate-conscious behaviors are actually to a person's advantage.[7] Drinking tap water is cheaper than bottled. Turning down the heat or air conditioning saves money. Eating vegetables is healthier and generally better for you than meat. Reusing and repurposing saves money too. The list goes on. You can't criticize these as elitist, can you?

Eliza: Under normal circumstances, no.

Hope: Okay! So, can we recommend that students adopt these lifestyle changes?

Do Your Carbon Emissions Make a Difference?

Eliza: (*reluctantly*) I still think that focusing on individual emissions misses the point. My emissions don't make a difference, and anyway climate change is mainly a structural problem. So, it feels like we're putting our emphasis in the wrong place.

Andrew: I'm confused. Do you think that your personal emissions do not, by themselves, make a difference to the climate problem?

Eliza: Maybe. It takes a huge volume of emissions globally to make a difference to the climate. The carbon dioxide I add to the atmosphere *myself* is tiny. Even my emissions across my whole lifetime aren't going to harm anyone by themselves.

Hope: Hmm. I'm not sure you're right on that one, Eliza. (*a thought strikes her, she pulls out a notepad and begins writing. Eliza and Andrew exchange surprised looks*) Let's take a stab at some numbers. Suppose we think about it this way. (*Hope continues writing*) The United

7 Schwenkenbecher, 2012.

States accounts for about one-fifth of annual global emissions, and there are about 300 million of us. So, dividing that out ... we get that the average American is responsible for 6.7×10^{-10} of global annual emissions.

Eliza: Or, in English, a very tiny fraction.

Hope and Andrew chuckle.

Hope: Still, if we're asking about individual responsibility, we should probably focus on lifetime emissions, because climate change is not a product of *annual* CO_2 emissions, but rather total emissions since the *industrial revolution*.

Eliza: So, what is the average American's contribution to total emissions over their lifetime?

Hope: Larger than you think. I'll spare you the mathy details,[8] but here's a general idea. Let's take an American that

---

8 Nolt, 2011b. Here's his full explanation of the calculation:

"[To calculate individual responsibility,] we will need to know how much of the total harm of climate change to attribute to the average American. I will here assume that harm is proportional to emissions. Current US emissions are about one-fifth of the world's. There are about 300 million Americans. This means that the average American is responsible for approximately $1/5 \times 1/300,000,000 = 1/1,500,000,000 = 6.7 \times 10^{-10}$ of current *annual* global emissions.

The total harm of global climate change will result, however, not only from this year's emissions, but from the total anthropogenic increase in atmospheric greenhouse gas concentrations that began with the industrial revolution. An individual can be complicit only in those greenhouse gas emissions that occur during her lifetime – not those that occur before she was born. But she is, I assume, complicit in emissions that occur during her lifetime. The lifetime of the average American is about eighty years. For the sake of definiteness (though there is no point in being too precise here), let us say that this average contemporary American's life begins in 1960 and ends in 2040.

From these dates, we can calculate the increase in the atmospheric concentration of carbon dioxide during her lifetime. The concentration at the beginning of her life in 1960 was 317 ppm. Today (late 2009) it is 388 ppm. Projecting the current growth rate of about 2 ppm/yr into the near future we obtain an estimate of 450 ppm for 2040. Thus total increase during her lifetime comes to $450 - 317 = 133$ ppm. The pre-industrial concentration of CO_2 was approximately 280 ppm (IPCC, 2007, p. 37). So before this average American's birth, the atmospheric CO_2 concentration had increased 37 ppm. During her lifetime it will increase an additional 133 ppm. *(Continued)*

lives between 1960 and 2040. She will likely be alive for 78% of the anthropogenic carbon emissions since the industrial revolution. That would make her responsible for about *one two-billionth* – I calculated 5.2×10^{-10} – of the greenhouse gases contributing to anthropogenic climate change by 2040.[9]

Eliza: Really?

Hope: (*grins*) My calculation was pretty back-of-the-envelope – simple arithmetic combined with a quick internet search – but I think it gets us in the right ballpark.

Eliza: (*examining the notebook*) Impressive! But doesn't this prove my point? If I'm only one two-billionth responsible for the release of greenhouse gases, that's practically the same as being not responsible at all.

Hope: Surprisingly, no. We tend to underestimate the amount of harm climate change will do. May I? (*Eliza slides the notebook back to Hope*) The IPCC estimates that current emissions will contribute to warming and sea-level rise for a millennium ... (*in a mutter, almost to herself*) ... that means about 100 billion people will be affected by anthropogenic climate change ... (*Eliza and Andrew lean over to look*) Suppose that only 4% of these people are harmed or killed, then that's still 4 billion people. ... If the average American is responsible for one two-billionth of that harm .... (*circles the answer and*

The total anthropogenic increase until the time of her death therefore is 37 + 133 = 170 ppm, of which about 78% will have occurred during her lifetime.

We noted above that the average American produces approximately 6.7×10^{-10} of current annual global emissions. Let us assume, as seems reasonable, that her portion of the annual emissions stays roughly the same over her lifetime. We have just seen that 78% of the total increase in atmospheric greenhouse gases to 2040 will have occurred during her lifetime. It follows, then, that her portion is about $6.7 \times 10^{-10} \times 0.78 = 5.2 \times 10^{-10}$ – that is, about one two-billionth of the greenhouse gases contributing to anthropogenic climate change by 2040" (Nolt, 2011b, pp. 7–8).

9 Nolt 2011b, p. 8.

spins the notebook around again) Got it! Our individual share is 1–2.

Andrew: Wow! Are you trying to say that the average American will, due to their individual lifetime emissions, be responsible for the suffering or death of 1–2 future people?[10]

Hope: Given reasonable assumptions, yep.[11]

Andrew: Well, that seems like an excellent reason to reduce your emissions, doesn't it?!

Eliza: I'm sorry; I'm not convinced.

Hope: Why not?

Eliza: The methodology is too simplistic.

Hope: In what way?

Eliza: Well, your calculations used *per capita* emissions to determine responsibility: the first thing you did was divide America's greenhouse gas emissions by its total population.

Hope: Is that a problem?

Eliza: Yes! It's far too simple. You started by saying that the United States accounts for about one-fifth of the world's emissions, right? So, you're talking about the total emissions *produced* in the United States.

Hope: Right.

Eliza: What about the emissions *consumed* by Americans?

Andrew: I'm not following.

---

10 Nolt, 2011b, p. 9.

11 Nolt, 2011b. Nolt wrote his paper to challenge the widespread assumption in the philosophical literature that individual emitting behaviors do not cause morally significant climate harms. His striking conclusion relies on three main assumptions: (1) individuals (or organizations reducible to individuals) are the only responsible agents for climate change; (2) equal-per-capita is a roughly accurate measure of the "average American's" emissions; and (3) that spread harm is morally comparable to concentrated harm (for a discussion of the difference between spread and concentrated harm, see Fragnière, 2018, pp. 653–656). All of these assumptions can be contested, but they aren't obviously false on their face. Accordingly, Nolt's paper successfully challenged those who claim that one individual's emissions do not cause morally significant harm to justify their position.

Eliza: There's a difference between consumption emissions and production emissions. Say it takes 12 tonnes of CO_2 to build a car. You want somebody to be responsible for those emissions, but who? You could say those who built the car, or the person who buys the car. It can't be both, though, because then you would be double-counting emissions.

Andrew: What do you mean?

Eliza: If you claim both the producer and the consumer are fully responsible for the 12 tonnes of emissions, then you would be saying that together they are responsible for 24 tonnes of emissions, even though only 12 tonnes were released in the atmosphere. You'd be double-counting.

Andrew: I see. I'd say the consumer is responsible. In the end, they are who the car is *for*.

Eliza: But simplistic equal-per-capita approach doesn't account for that.

Hope: Point taken. Still, my calculations were in many ways conservative. The per-capita estimates I used were based on production emissions as Eliza said. But we import more than we export, and a lot of the goods we consume come from other places, especially China. So, our consumption emissions will *actually be higher* than our production emissions! I think the number is about 6% higher.[12] Meanwhile, China's consumption emissions are about 15% *lower* than their production emissions.[13] In other words, if anything, I underestimated American responsibility.

Eliza: I see.

Hope: I may also be underestimating it in other ways. For instance, I assumed only 4% of future people will be harmed by climate change. If you raised that number, that could easily make up for my overstating the average

12 Ritchie, 2019.
13 Ritchie, 2019.

American's responsibility in the crude calculation – if I did overstate it; arguably, I played it down. Anyway, I'm only trying to give a rough benchmark and surely it is good enough for our purposes.

Eliza: Fair enough. Still, surely the wealthy are disproportionately responsible for consumption emissions.[14] Given that, even per capita *consumption* emissions don't tell the whole story.[15] Your methodology *implicitly* makes poorer Americans take responsibility for the emissions of rich Americans.

Andrew: Eliza's right; I think we need to be really careful about that.

Structural Emissions

Eliza: (*reluctantly*) There's a bigger issue behind all this. I still think that focusing on individual emissions misses the point. Climate change is mainly a structural problem.

Andrew: What do you mean?

Eliza: Climate change does not primarily result from individual wrongdoing, but instead from a global economy that relies on the burning of fossil fuels for its basic operations. It's not my fault that when I heat my home, or drive to work, I emit greenhouse gases. Even more importantly, nothing will *change* by *my* turning down the thermostat or taking the bus. Climate change will keep happening until the global energy infrastructure changes.[16] *That's* why climate change is a structural problem. And *that's* why critiquing individual carbon footprints largely misses the point.

Hope: Interesting. Say more.

---

14 Oswald, Owen, & Steinberger, 2020.
15 For more discussion of the practical and ethical complications of carbon accounting, see Steininger, *et al.*, 2016.
16 Hale, 2011; Aufrecht, 2011; Weisbach 2016.

Eliza: Well, your various calculations simply assumed that it is *individual consumers* who should be held responsible for emissions.[17] But that seems naïve to me. What about the responsibility of corporations and governments? You seem to be neglecting how the world actually works!

Andrew: Can you give an example?

Eliza: Sure. Let's say I use a lot of electricity at my house, and thereby accumulate a large carbon footprint.

Andrew: You should use less!

Eliza: Perhaps; but why does the responsibility fall on *me*? I didn't decide for my electricity to be generated through coal or natural gas. The utility company and maybe the government did.[18] Why aren't they the one's responsible? I would have preferred hydroelectric, wind, or solar. But they gave me no other choice.

Andrew: Okay. I agree the utility company is *partially* responsible.

Eliza: I think they're *mostly* responsible! My worry is that, in assuming that responsibility only falls on individuals, Hope's back-of-the-envelope calculation ignored the huge role corporations and other big actors play in *bringing it about* that individuals *have* such large carbon footprints.

Andrew: Good point. Corporations are important: more than a third of all carbon emissions since 1965 can be attributed to the largest 20 fossil fuel companies.[19] We shouldn't ignore that.

Eliza: One risk here is that getting obsessed by individual choices is exactly what massive corporations want, as a way of deflecting blame and distracting attention. Some even say that the whole idea of individual carbon footprints was originally popularized by fossil fuel companies.[20] It

17 Hartzell, 2011.
18 Hartzell, 2011, p. 16.
19 Heede, 2019.
20 Doyle, 2011.

is so convenient for them to sap the energy of activists by encouraging people to focus on things that won't make a big difference in the long run. While we cajole people about greening their personal lives, we're ignoring the fact that corporate behavior is rendering those efforts largely irrelevant.[21] We need major, concerted *political* action. At the very least, individual emissions reductions are far from *sufficient* for solving the problem.[22]

Hope: I wonder if you're being too harsh. Maybe the corporations are just trying to confront things from the demand side, as well as the supply side. You know, encouraging people to help them to reduce emissions.

Andrew: Hope! Sometimes your faith in people is endearing; but in this case I think you're really being naïve!

Eliza: (*exasperated*) Whatever. Anyway, regardless of intentions, can't you see now why I think climate change is mostly a structural issue, not an individual one? I'm not saying that people aren't responsible for any of their own emissions. But decisions by others – especially corporations and governments – heavily structure the options available to individuals. That's why I think focusing on people's carbon footprints misses the point.

Andrew: I see where you're coming from, Eliza. But I don't see why it's an either/or. We certainly need to be taking on corporate power by demanding regulation and a tax on carbon. I've spent countless evenings and weekends waving signs demanding just that. But we *also* need to be reducing our carbon footprints.

Eliza: In some ideal world, maybe. But I worry that *in practice* it is more of an "either/or" than you think. I'm not sure it is productive to try to have it all. We have only limited amounts of energy, time, and attention. Maybe it is worth doing some things at the individual level, but

21 Lukacs, 2017.
22 Cuomo, 2011.

I think we shouldn't devote too much of our precious energy to greening our personal lives when such reductions don't ultimately make a difference.

Andrew: You haven't shown that yet. You've only suggested that they don't make the *biggest* difference.

Tipping Points and Risks of Severe Harm

Eliza: Really? Isn't there a good case for *no* difference? Unless we decarbonize our economy and everyone else stops emitting, the same amount of deaths will happen. Surely, *ought implies can*: it can only be true that I *ought* to reduce my emissions because they are causing harm if it's also true that I *can* prevent that harm by reducing my emissions.[23] Otherwise, you're just asking for a pointless sacrifice. If I never emit carbon, dangerous climate change will happen all the same. My "share" of a massive collective harm – what Hope calculated as my emissions harming 1–2 future people – ignores this point, and so is morally misleading.[24]

Hope: I'm not sure. For one thing, there's good reason to think that there are certain "tipping points" of carbon in the atmosphere that, once passed, will cause extreme harm.[25] Your emissions might send us over one of those tipping points. Then, *you* might make *all* the difference!

Eliza: But it is really unlikely that *my* emissions would prove decisive.

Hope: Hmm. Even if the risk of my harming someone by pushing total emissions over a threshold of catastrophic harm is small, shouldn't I eliminate that risk because the harm

---

23 Seager, Selinger, & Spierre, 2011, p. 40; Baatz, 2014, p. 9.
24 Fragnière, 2018, p. 654.
25 Lenton et al., 2019; Hiller, 2011.

is so large? We're talking about genuine catastrophe here.[26]

Andrew: Exactly. It is wrong to drive drunk even if you luck out and don't harm anyone, because driving drunk puts other people at risk of severe harm.

Eliza: But climate change is not like drunk driving! When someone drives drunk and kills someone, it is easy to identify the perpetrator and the victim.[27] But in the climate case the system is going to be way too complex for that.

Andrew: Okay. Maybe it is hard to *trace* the harm in the climate case; nevertheless, that isn't the same as your emissions actually being harmless.[28] Plus, even if your emissions don't *in fact* cause harm, you shouldn't take the risk that they will.

Eliza: I'm worried that your view has become way too demanding. If that were true, basically everything I do violates a moral obligation![29] When I *breathe* I increase the amount of CO_2 in the air. If I shouldn't increase the risk of harm to others at all, then I shouldn't breathe! Maybe I shouldn't exercise either, since I breathe more heavily then!

Hope: Well, let's be fair. Minimizing the risk of climate catastrophe constitutes a significant moral reason not to emit, but that doesn't mean that it's wrong to emit all-things-considered in every case.[30] Obviously, you'll die if you don't breathe, and exercise is important for one's health. So, these things morally justify you breathing, and exercising as well.

---

26 Broome, 2019, p. 118.
27 Sinnott-Armstrong, 2005, p. 302.
28 Broome, 2019., p. 116.
29 Sinnott-Armstrong, 2005, p. 302.
30 Obst, a.

Eliza: But what if there's a risk – even a tiny risk – that my morning run will push us over a threshold into climate catastrophe ...?

Andrew: Sure, there may be real questions at the margins. But can't we at least agree that there are a whole host of activities that should normally be off-limits? Like, you ought not to go driving around the countryside on a Sunday afternoon in a big, gas-guzzling SUV just for the sheer fun of it. Joy-guzzling is blatantly irresponsible.

Eliza: Maybe; but why? You haven't shown that going for a joyride *significantly* increases the chance of harm to future people. If my green-living won't prevent harm in most cases, and everyone else is emitting greenhouse gases, I'm not seeing that there's a significant moral reason for me to stop emitting.[31] Instead, I should push to make fossil fuels illegal, which will actually make a difference.

Agreements and Social Norms

Hope: I think there might be some kind of fallacy here. If all of us stopped, then we could stop the deaths, and not just one or two but potentially billions. Yet, if each person thinks like you about the small difference they make alone, it becomes a self-fulfilling prophesy that no one acts, or only a very few. And that just seems too convenient.

Andrew: Yeah, what Hope said. Let's try an analogy. I'm guessing you don't plan to vote either?

Eliza: What do you mean? I'm a very loyal voter. My main point is that we should be working collectively to get our

---

31 Sinnott-Armstrong, 2005; Kingston & Sinnott-Armstrong, 2018 defend roughly this view. For a lengthy discussion of the ethics of expected harm and individual emissions, see Cullity, 2019.

government to take climate action.[32] It is one of the best ways to do that! It's the least you can do.[33]

Andrew: So, you don't think voting is basically pointless? After all, it's very unlikely that your vote will make the difference in an election. And it is onerous, showing up at the polling station, standing in line, all of that.

Eliza: You're trying to turn my arguments in the climate case back on me.

Andrew: I'm just pointing out that if you claim that we don't have a duty to reduce emissions because they don't make a difference, it seems like similar reasoning would mean we don't have a duty to vote.

Eliza: Voting is different. The likelihood of your vote being the one that matters is much higher than the chance your emissions cause harm.

Andrew: I think you are exaggerating the likelihood that your vote will make a difference. Studies have shown that the chances of one's vote making a difference is infinitesimal if one has good reason to think that one candidate has a non-trivial lead.[34]

Eliza: What, so now you're arguing that I shouldn't vote?

Andrew: No; I think you should! Even in a state where the candidate you don't want is probably going to win and when your vote is almost certainly not going to make a difference.

Eliza: I agree. Not voting is free-riding: reaping the benefits of other people voting, but not voting oneself.

Andrew: I think the same thing about people who won't act on climate, who don't let it affect their political behavior or their personal choices. They are free-riding, hoping that the rest of us will deal with it.

---

32 Sinnott-Armstrong, 2005, see annotated bibliography; Young, 2011, pp. 165–170; Cripps, 2013, p. 142.
33 Maltais, 2013, see annotated bibliography. Also, Woodson and Gardiner 2019.
34 Budolfson, 2019, p. 1718. However, for a rejoinder, see Barnett, 2020.

Eliza: *(thinking)* I suppose I think the main difference is that there's a social norm or collective agreement that we should vote. We have decided that voting is best for everyone and that everyone should vote, so refusing to vote would be free-riding on others' efforts. But I'm not free-riding by emitting, because there is no collective agreement yet that we should all be carbon neutral. Unilateral reductions won't stop the problem, and so aren't required. Rather, our obligation as individuals is to try to establish a collective agreement against emitting carbon.[35]

Hope: What about the collective agreements we already have? In 1992 in Rio, countries agreed to prevent "dangerous anthropogenic interference" with the climate system under the UNFCCC. In Paris, they agreed that this means holding the increase in global temperature to "well below" 2 degrees, while "pursuing efforts" to limit it to 1.5 degrees. So, why not say that everyone – countries, businesses, individuals – has a duty to bring their behavior in line with those agreements? To be Rio-compliant and Paris-compliant? Don't we at least have good reasons to cooperate?

Eliza: Hmm. I hadn't thought about it that way. Still, those agreements are at the international level. Also, Paris in particular is legally weak, and there's no real enforcement.

Hope: Maybe so. But that might not be the central issue. After all, don't you think that there are certain behaviors we ought to *expect* from people, certain virtues they think they are obliged to have? This applies to a whole range of things, right? People shouldn't loot even if others are doing it; they shouldn't dump their trash in the neighbor's yards even if others have already done it; they

---

35 Johnson, 2003, p. 283.

shouldn't steal from others even if their victims are rich and could afford it?

Eliza: Agreed. Ethics isn't just about consequences, it's also about what you do and who you are.[36] But, again, there are collective agreements not to loot or litter – that's why it's wrong to do them. You'd be free-riding. By contrast, there isn't a collective agreement not to emit, not *within* our society anyway.[37]

Hope: Don't you think individuals' personal efforts to reduce emissions may help galvanize support for collective agreements at the national or local levels? People learn moral norms from one another. Your excessive fossil-fuel use may send a message to others that unsustainable emitting is acceptable.[38] If they believe that, why would they support a carbon tax or any other collective climate agreement? Excessive emitting seems counterproductive at best, and morally blameworthy at worst.

Eliza: Maybe; but I'd like to see more evidence. Perhaps I can do a better job galvanizing support by concentrating on acting politically, not individually. If I go all in on reducing my own emissions – wearing the hair shirt – that might isolate me, it may even alienate a lot of people.

Andrew looks annoyed.

Eliza: Anyway, now I'm wondering whether what you really think is that I have an obligation to dramatically cut my emissions to encourage others no matter what those others are actually likely to do. That would be very uncompromising.

---

36   The latter idea is a defining feature of the approach known as virtue ethics. For a general survey, see Hursthouse & Pettigrove, 2016. For a discussion of why virtue ethics may be especially appropriate for environmental issues, see Sandler, 2010.

37   Johnson, 2003, p. 283.

38   Hourdequin, 2010, p. 453.

Andrew: (*hesitates*) Well, yes. I think that being uncompromising is the best way – perhaps the only way – to produce the right outcome.[39]

Eliza: But surely that would be way too demanding! What you're proposing would make many of my life projects – the things that make my life worthwhile – impossible. And you demand this of me even if there's not much chance that it will make a difference to other people or the improving the world.

Andrew: Yeah, it sucks. But climate change sucks, and meeting our moral obligations isn't easy. Also, we in the rich, industrialized world have to remember that, while *we* have been living good lives, many on the planet have not had that opportunity.

Fair Shares

Hope: Maybe we can be a bit more nuanced. Eliza is surely right that there are issues of fairness lurking here. Suppose we begin differently, with the thought that there is still an amount that current people can emit without causing catastrophic climate change. Maybe each of us is entitled to a fair share of that carbon budget. Taking more than one's fair share, though, both contributes to a harmful activity, and deprives others of their fair share by using up what's left.[40]

Eliza: How do we figure out an individual's personal fair share?

Hope: Obviously, that's a big question. But we might get a rough idea – an initial benchmark – if we see what an equal share of the remaining global emissions budget might be.

Eliza: Okay. What's that?

---

39 Jamieson, 2007.
40 Baatz, 2014, see annotated bibliography.

Andrew: Suppose we took the total remaining budget to keep below 2°C and divided it by the number of people. Other things being equal, there's a presumption that most people shouldn't be emitting more than that. In fact, that's pretty generous because 2 degrees is probably too high a target, and bad things are happening on the ground already. So, arguably we should be cutting much more quickly.

Eliza: Fine. On this "generous" reading, what's my fair share?

Andrew: Well, it turns out that the sustainable level would be about 2–3 tons of carbon dioxide per person per year based on 2015 numbers.[41] So, that's the benchmark for a fair share.

Hope: I'm not sure I even want to know, but how close are we to that level?

Andrew: (*wincing*) Not remotely close. The current global average is more like 6.2.

Eliza: So, we're way over.

Andrew: It is actually worse than that: 6.2 is the global average. The American average is about 16; by comparison the United Kingdom's is 5.8, South Africa's 8.5, and Vietnam's 2.1.[42]

Eliza: I get it. Most places are over the 2–3 ton benchmark; but we Americans are way over.

Hope: Ah. So, as Americans we probably shouldn't be complaining that cutting back is unfair *to us*.

Andrew: Of course, we must remember that not all Americans are the same, nor all Africans! I read recently that the richest 1% of Americans – 3 million people – emit 318.3 tons of carbon per person per year on average, and the richest Saudis and Singaporeans about 250.[43]

---

41   Baatz, 2014.
42   World Bank, 2016.
43   Chancel & Piketty, 2015, p. 29.

Hope: Wow. That's a huge difference from the global average of 6.2.

Andrew: Yes. By contrast, the global poor tend to emit very little. The averages for Honduras, Mozambique, and Rwanda are around 0.1.

Eliza: I agree that those are stunning comparisons. Still, they underscore my basic point from earlier. Notice that those countries with really low per capita numbers are places where the infrastructure is not yet infused with carbon. In the United States, you probably can't get your carbon footprint down to your fair share unless you head for the hills and go "off grid." Maybe the billionaires and multimillionaires can do that. But it's ridiculous to tell many Americans to consume less energy without addressing the structural issues; many people are just using the energy they need to get by. They'll tell you that while you're worrying about the end of the world, they're worrying about getting to the end of the month.[44]

Andrew: How about this? You're right, there's a difference between necessary emissions and others that feel more optional. So, I'm not going to argue that Americans have to give up their *subsistence* emissions: for example, those that are needed just to eat, sleep, and stay warm. But *luxury* emissions are a different matter.[45]

Eliza: I agree that the richest 1% in the United States can't claim that most of their emissions are necessary.

Andrew: And the same can be said of many of the emissions of the rest! That's why most of us need to be drastically reducing our individual emissions.

Eliza: But, as I've said, many times certain emissions might be difficult to avoid. They are structural – built into the system.[46]

---

44 Goodman, 2019.
45 Shue, 1993. For further discussion, see Dialogue 4.
46 Aufrecht, 2011.

Andrew: Yeah, it might be difficult. But, as I said, it shouldn't
          be surprising that the prospect of something as morally
          bad as catastrophic climate change demands a lot of the
          individual.

Part 2: The Extent of Responsibility

Hope: I feel like we've reached an impasse. I wonder if there's
          some middle ground. I partially agree with Andrew,
          but I also somewhat agree with you, Eliza. It's not rea-
          sonable to demand that individuals emit *only* what is
          required for subsistence, or even only when necessary to
          live a very minimally decent life.

Eliza: Do you think it might be permissible to fly for work, or
          for a vacation?

Hope: Probably; at least sometimes. Still, everyone can be
          doing *something* to live their personal lives more sus-
          tainably. Perhaps we should think of it like exercise. If
          it's not causing you any pain, you're probably not doing
          enough; but if you're exerting yourself to the point of
          severe discomfort, you're probably doing *too* much.[47]
          Never flying for leisure, at least to me, is excessive in
          that latter sense.

Andrew: Wait! How could you say that, Hope? I thought you
          agreed that by emitting we increase the risk of climate
          catastrophe and future harm? And you also seemed to
          agree that many Americans were emitting more than
          their fair share? Doesn't emitting more than the mini-
          mum make us responsible for serious harms in the
          future? You can't justify causing serious harms to others
          merely on the grounds that it will make your own life
          better.

Hope: I guess I'm not sure I'm responsible for the harms that my
          emissions might cause *in the same way* I'm responsible

---

47   Raterman, 2012. See annotated bibliography.

in normal cases, like if I punch someone in the face. The moral reasons at stake seem less weighty in the climate case, and so more likely to permit emitting than punching.[48] But my main point is that I agree with Eliza that the climate problem's structural nature means that fighting for collective change is more important.

Andrew: Sounds like a cop-out to me.

Comparing Lives

Hope: I agree there's a risk of that, and we should be on our guard. I wonder if it would be helpful to compare two lives.[49] In one of them, the person, call her Serena, does her very best to reduce her own emissions. She doesn't fly, becomes vegan, gives up the car, eats local, vacations close to home if she vacations at all, installs solar panels ... all that stuff. Serena is an environmentalist, but her focus is on the actions she takes as an individual. She thinks of herself as living an ecologically good life, striving toward the kind of life consistent with an environmental utopia, even though ours is not such a society.

Andrew: That's basically who I want to be.

Hope: I honestly believe you deserve praise for doing so. Still, consider a different life. This person – call her Grace – also cares deeply about climate change; in fact, she organizes much of her life around fighting against it. But when she does this, she doesn't worry too much about her own footprint. She travels a lot, to participate in rallies, to educate people about climate, and is always looking for opportunities to engage people at every level, whether in business, or politics, or government. In other ways, she also lives a fairly normal life by the standards of her

48   Fragniere, 2018, p. 658.
49   This rest of this section draws on Gardiner, in preparation.

society. She drives to work, eats meat, takes vacations, flies to conferences, and so on.

Andrew: So, she doesn't care about her own emissions at all? (*skeptically*)

Hope: (*sighs*) She does care. She tries not to do egregious things. She's not driving a Hummer or investing in drag racing, or anything like that. And she tries not to go on massive trips unless there's a good, climate-related reason to do so. She also buys offsets for some of her activities, though she's not fastidious about it.

Andrew: Grace sounds like a hypocrite. In fighting for collective change while not doing that much to reduce her personal contribution, her behavior contradicts her beliefs. It seems clear that she doesn't really hold those beliefs, or is insufficiently committed to them. She's all "do as I say, not as I do."

Eliza: I don't think that's fair. She is doing *a lot*.

Hope: I'm not sure she's a hypocrite. Suppose that when people accuse of her hypocrisy, she responds by saying she isn't perfect, but she's doing a lot for the climate. She's dedicating her life to climate action, organizing much of it around that project. In that way, she's doing more than most of her peers, including some of those who go vegan and stop flying. Do we really want to say that she's a hypocrite? Or even that she's living a worse life than the stay-at-home activist, even though otherwise she lives the life she would have chosen anyway?

Eliza: This is what I meant earlier by the focus on individual emissions feeling condescending. It's like a purity test.

Hope: That bothers me too. All the same, let's say Grace is not totally comfortable with her choices and lifestyle. She worries about whether she's succumbing to convenience and self-serving reasoning.

Andrew: You mean like the moral corruption thing you mention all the time?

Hope: Exactly.

Andrew: So, why isn't she corrupt?

Hope: Ultimately, she thinks she's striking a reasonable balance. Anyway, much of the criticism is coming from people who are hostile to the climate action, whose opinions she doesn't give much weight anyway. Of course, she respects those who want to be pure, even supports them in some ways. And she thinks there is a real dispute here. Still, she believes that the solutions ultimately have to be political, so that's where she's putting her energy. The other stuff matters, to be sure, but it's not at the heart of what needs to be done, at least it's not the best use of her time and energy.

Andrew: (*rolling eyes*) I still think it smacks of moral corruption. Anyway, it is a false dichotomy. She can do both! Fight for collective change but try as hard as she can not to contribute personally.

Hope: Maybe. But I'm worried about something here. It's like the "mommy myth"; you know, the one where the ideal woman has the perfect career, the perfect family, the perfect body, all while being the perfect wife and mother and eating the perfect diet. She is perfect in everything, and those women who fall short of this ideal deserve blame.

Eliza: You're talking about the unrealistic pressures placed on women and girls in our society, the pressure to be and do everything. Not only is this myth unrealistic, but it is also oppressive to women. We get put in a "no-win" situation and then criticized if and when we don't quite match up, whenever we "don't quite make the grade." But the game is rigged against us.

Hope: Yeah. I think that the right thing to do is not even to play that game. Even by trying, you lose.

Andrew: I see. So, are you saying that climate activism can be like that too? That we can be tricked into playing the wrong game?

Hope: Yes. Personally, I don't like the "hypocrisy" game. I'm not sure what the rules are. I'm also worried that they're often being manipulated, so that things certainly can't end well for me, and probably can't end well for the movement either, no matter what any of us do. I can't help but notice that it suits the deniers and obstructionists perfectly if we tie ourselves in knots denouncing each other about purity vs politics. You know – the circular firing squad.

Andrew: I'm against firing squads! I also agree that sometimes the hypocrisy game can be toxic and that nobody's perfect. But I still think Grace is probably a hypocrite. Do you really think that she isn't doing something morally wrong?

Hope: What do you think she is doing wrong?

Andrew: Well, you say that she doesn't do "egregious" emitting, but that she doesn't worry much about her own footprint. That seems wrong. If she is dedicated to avoiding dangerous climate change, she should care about how her behaviors contribute to the problem. She should think hard about them. She should show integrity.

Hope: Isn't that a little unfair? I said Grace cares some. She just doesn't dwell on it. She's "thought hard," as you put it, about how to live her life in the face of the climate threat. But she's decided not to put much energy into personal emissions; some, but not much.

Andrew: Let's take a positive spin on this. Imagine a third person – Rashida – who both dedicates her life to fighting for climate justice at a political level – by activism, by voting, through protest – but also takes great care to reduce her own emissions as much as possible. I think they have far more integrity than somebody who does one but not both.

Hope: Do you think that one has a moral obligation to have integrity?

Andrew: I do! Everyone ought to work to harmonize their commitments at various levels and achieve a life in which their commitments are embodied not only in a single sphere, but in all the spheres they inhabit.[50] One should seek a life of internal harmony, where one's political goals harmonize with one's private behavior.

Eliza: So, you think that a climate activist should *never* fly or drive?

Andrew: I'm not saying that. Rather, I want to deny that this is a question of *purity* vs politics in the first place. Maybe sometimes there can be good moral reasons to fly or drive. But I think that a climate activist with integrity must always be thinking about their carbon emissions and doing their best to reduce them.

Eliza: That's a lot of thinking. I'm not sure it's reasonable or productive.

Andrew: It is productive! Reducing personal emissions facilitates political action, because it helps activists identify areas where it proves very difficult to reduce emissions in their own lives. That can inspire them to push for specific policies.[51] If I find that there is no reasonable way for me to get to work by bus – it's an hour and a half, with three transfers along the way – this can prompt me to push for better public transportation.

Eliza: What do you think of Greta Thunberg, and her decision to sail across the Atlantic rather than fly?

Andrew: I think Greta's decision to sail was a morally good one and sets a good example. In fact, that's another reason why climate activists should care about their carbon footprint. As Hope alluded to earlier, by going vegan,

---

50 Hourdequin, 2010, p. 449. See annotated bibliography.
51 Many thanks to Marion Hourdequin for raising this point in her commentary on an earlier version of this dialogue.

or rarely flying, you send a message to others that they should do the same. So, the impact of reducing one's carbon footprint goes beyond the immediate benefits of that carbon not being released into the atmosphere. You can affect moral reform through your actions by serving as an inspirational model.[52]

Eliza: I wonder. Couldn't you say that Greta sets a *bad* example with such an extreme choice? After all, not everyone is famous enough to get offered free passage across the Atlantic!

Andrew: You make it sound like a pleasure cruise! The boat she took had no shower, toilet, cooking facilities, or proper beds. She made significant sacrifices to reduce her carbon footprint, which sets a good example.

Eliza: Fair enough. Still, that kind of sacrifice might *discourage* people from reducing their carbon footprints or supporting a decarbonized economy. Moreover, as you've admitted, there may be good reasons to fly. Doesn't Greta's behavior send the message that flying shouldn't happen at all in a climate-conscious society?

Hope: There's something to that. Do you know about Luisa Neubauer? She's another young climate activist who chooses to fly in service of her climate activism, and now she's getting called a hypocrite by some climate activists.

Eliza: I think Greta's choice to sail across the ocean rather than fly encourages this counterproductive flight shaming. This is exactly why I don't like all this focus on individual emissions. Like Hope said, there's no winning.

Hope: I have another concern. Some claim that those working on climate – scientists, academics, activists – shouldn't be flying. But I have to say that stance worries me. If anything, I think that those working for change are the

52   Hourdequin, 2010, p. 453.

very last people who should stay home. Maybe they even have a *duty* to fly if needed, and we have a duty to support them.

Eliza: Yeah. If their showing up at an important meeting might make a difference, surely they should go?

Hope: I might be in favor of the rest of us devoting some of our fair share of emissions to the activists so they can keep doing what they're doing. After all, we don't want climate skeptics to be the only ones flying around raising awareness; we need climate activists at the table, and flying is often necessary for that.

Andrew: I see what you're saying, but it also seems very convenient. C'mon you two! You're just the kind of climate activists you are describing! It sounds like you wish to keep flying, and you're rationalizing it.

Hope: I agree that it is a risk. Maybe people will see it that way. So be it. I can live with that. I tend to think they're not the audience that matters in the end anyway.

Andrew: What do you mean?

Hope: I think its future generations and the rest of humanity that matters. Can I justify my actions to them? I think I can; but I'm also willing to take the risk that I'm wrong. I don't take that lightly; if I am wrong, I'll deserve their censure. Still, I think they will be somewhat forgiving. At least, I think they will see someone going out and giving it an honest try. I'm sure they'll respect you as well for trying to emit as little carbon as possible. I'm just not convinced that focusing on one's own carbon footprint is the only or best approach. They might even prefer my strategy to yours.

Andrew: I hear you. Maybe we don't disagree all that much.

Offsetting

Hope: It's worth remembering that reducing consumption isn't the only way to reduce your carbon footprint.

Andrew: What do you mean?

Hope: You can offset your emissions.

Andrew: You mean like pay people to plant trees that will suck CO_2 out of the atmosphere to compensate for what you emitted?

Hope: Exactly.

Andrew: I've always been suspicious of offsetting programs. Aren't they like rich churchgoers in Mediaeval times buying "indulgences" to "offset" their sins?[53] Or a rich hiker throwing his beer can in the Grand Canyon and saying he'll just pay the fine to "offset" the pollution?[54] Just don't do the bad thing in the first place!

Hope: I don't see it like that at all!

Andrew: Why not?

Hope: Well, first of all, offsets are currently very cheap![55] They are not reserved for the rich. Second, remember that the harm done by your emitting CO_2 comes about through the effect it has on the global concentration of green-house gases.

Andrew: So?

Hope: If you don't *add* to the global concentration, you don't cause harm. If you emit at one place, but prevent an equal quantity of emissions at another place, you don't change the global concentration, and so do no harm.[56] Offsetting your emissions is more like not sinning at all.

Andrew: I guess I never thought about it like that. I see your point; but isn't there something wrong in polluting regardless of the harm it causes? Consider that rich guy throwing his beer can into the Grand Canyon and paying someone to pick it up. I can't help but think that it reflects poorly

---

53   Goodin, 2010.

54   Sandel, 2005.

55   Some estimate that most individuals can offset their emissions for only $20 per month. See Thorpe, 2019.

56   Broome 2016, p. 159.

on his moral character.[57] What sort of person would do that? Same goes for the person who emits carbon wantonly because they know they can afford offsets.

Hope: Huh. I never thought about it like that either! I guess we're learning from each other. Still, I don't think avoiding harm shows a bad character.

Eliza: Something else worries me. Suppose that, for the sake of argument, I grant you that our carbon emissions are harmful. There are a million other harmful things we do every day. I contribute to pollution when I buy plastic, to sweatshops when I buy clothes, to corporate exploitation when I buy basically anything at all.

Andrew: Are you asking why you should reduce your carbon emissions – either directly or through offsetting – but not these other things? My answer's simple: you should reduce them all! You know I don't shy away from being demanding!

Eliza: I know! I'm not sure that's a virtue, though. I guess that, while I agree it's wrong to do *nothing at all* to withdraw myself from injustice and harm, I think that as long as I'm doing something, I'm fulfilling my moral obligations.[58]

Hope: I'm sympathetic. Still, I think there's something to be said for prioritizing reducing your carbon footprint. Climate change threatens great harm for hundreds, if not thousands, of years. It seems to me I have a special obligation not to contribute to that harm. Plus, I can "offset" my carbon footprint in a much more reliable way than I can "offset" my contributions to sweatshops. Honestly, I'm not even sure what "offsetting" my contributions to sweatshops would even involve.

---

57 Hill, 1983.
58 Lawford-Smith, 2016, pp. 140–141. For a more general discussion of moral responsibility in a world of ubiquitous injustice, see Lichtenberg, 2010.

Andrew: I have a practical worry. I've heard many carbon offsetting strategies are not effective. One thing is that the carbon in fossil fuels is much more secure in the ground than it is sequestered in a forest. The forest can burn down![59] Another thing is that you're assuming the money you give to offsetting programs is actually used to sequester carbon. Sometimes we're "saving" forests that were not threatened anyway![60] And some of the supposedly "non-profit" organizations take about 20% of what people contribute as running costs.[61] There are plenty of opportunities for shenanigans in the offsetting business.[62]

Hope: That is troubling.

Eliza: (*thoughtfully*) Hmm. That makes me wonder. If existing offsetting programs are far from perfect, maybe we have a duty to create better ones? And in the meantime, there might still be some role for the best programs – even if we lose 20% to running costs, 80% of the funds going to emissions reductions sounds worth doing.

Andrew: I'm still not sure. Given the problems – practical and moral – I think it's probably better to reduce one's consumption. Offsetting seems to me an easy pathway into moral corruption.

Hope: I wasn't suggesting that offsetting was some kind of silver bullet.

---

59 Choi-Schagrin, 2021.

60 For example, Simonet *et al.* found that 37% of REDD (Reducing Emissions from Deforestation and forest Degradation) projects overlapped with already protected areas (Simonet, *et al.*, 2015, pp. 17–18). See also Elgin, 2020.

61 Vidal 2019.

62 John Broome, a philosophical advocate of offsets, grants as much (Broome, 2012, p. 95). However, Broome may be overly optimistic about the prospect of differentiating ineffective from effective offsetting programs (e.g., Sagoff 2014, p. 196). Notably, Kyoto's Clean Development Mechanism (CDM), which granted emissions credits to countries that could demonstrate emission-reductions elsewhere, was heavily criticized for "numerous project activities and methodologies that do not fulfil basic environmental integrity and sustainable development criteria" (CDM-Watch, n.d.).

Eliza: I think we should also be aware that some emissions will not be avoided. Suppose a student needs to fly for a family funeral; at least there is a chance they can mitigate the harm by offsetting.

Hope: Okay. For the recommendations, perhaps we should just lay out the issues, so people can decide for themselves?[63]

Climate Anarchism

Eliza: Sure. But something still bothers me about the fact that focusing on personal behavior lets corporations and governments off the hook.[64] To me, too much emphasis on the individualist approach seems to embody a bad kind of political viewpoint, one which most of us would reject under normal circumstances, and which seems especially unlikely to work here. It looks like climate *anarchism*.[65]

Andrew: What do you mean?

Eliza: Anarchism is a political philosophy which says that everything should be done voluntarily. Anarchists reject conventional institutions, like governments, laws, and police forces, since they involve coercion. They encourage action at the individual level, or when cooperation is required, through loose, voluntary associations.

Hope: Sounds pretty utopian to me.

Andrew looks grumpy, but remains silent.

Eliza: To me as well, especially in the climate case. In fact, one reason I'm going to law school in the first place is that I

---

63 For a general overview of the practical and moral dimensions of carbon offsetting, see Hyams & Fawcett, 2013 (see, also, the annotated bibliography).

64 For a discussion of the ethics of negative emissions more generally, see Dialogue 5.

65 Gardiner, 2019a, pp. 207–208. Anarchism might be defined as "a political theory advocating the abolition of hierarchical government and the organization of society on a voluntary, cooperative basis without recourse to force or compulsion" (Lexico).

think we need laws and institutions to get our way out of this. I don't think a grand crusade cajoling people into giving up flying, cars, and having babies is going to work.

Hope: Do you mean the usual thing about the revolution taking too long, or taking way too many evenings?

Eliza: Not really. I won't explain all the reasons why I think anarchism isn't a great political philosophy.[66] My main gripe here is that because climate change is a genuinely global, seriously intergenerational problem, with too many perverse incentives; people won't comply without laws and institutions to hold them accountable. If we leave people to their own devices, we're doomed. Given that, why focus on the individual level?

Hope: I mostly agree. A political solution is needed. But I also think that support for political change needs to be built from the ground up, at the individual level.

Eliza: (*smiling*) So, you think that you've got to turn the whole country into a bunch of eco-loving vegans like Andrew before we can act?

Andrew: (*grouchily*) I think the chances of success will be better if we shift in that direction. We need to change people's minds and behaviors before we can effectively change the law. We'll also get a better set of people that way!

Eliza: Can't you see the political problem? You risk turning the climate fight into one about lifestyles, and for some that's about identity. I just want to focus on stopping climate change.

Anti-Capitalism

Andrew: Wait, I thought you were all about climate being a structural problem? Don't you think that consumerism and

---

66 Anarchism is not a popular view in political philosophy, but does have some staunch philosophical advocates. See Fiala, 2017, for an overview of anarchism and some standard objections and replies.

capitalism are at the heart of it? What about the whole global system, infested as it is with racism, sexism, massive inequality, and the legacy of colonialism? Surely, you're not denying the basic politics of the situation.

Eliza: I agree that there are serious problems out there. But I'm not convinced that a global revolution is needed just to fix climate change. And I don't think most people are ready for all that anyway. For one thing, if we tried to make fundamental systemic changes, we would have to agree on which changes those would be – just saying "capitalism must go" is clearly not a plan. For another, I'm not sure capitalism itself is the key problem. Maybe the system just needs better regulation.

Andrew: Pah! Your argument seems somewhat selective. Deep decarbonization of our society is *already* a fundamental change, and yet we recognize that justice *demands* it – and quickly.

Eliza: What about the fact that fundamental systemic changes take time, and require iteration and testing?[67] We need to act now, but the uncertainty associated with a global anti-capitalist revolution would require agreement on where we are aiming, and time to test new systems. And time is something we don't have.

Andrew: Malcolm X once said that "you can't have capitalism without racism."[68] I agree; I also think we can't have capitalism without climate injustice; so, climate justice *requires* we move on from capitalism.

Eliza: We should confront environmental racism and climate change at the same time. I just don't accept that overthrowing capitalism is *needed* to end racism. To be honest, I don't want to live in a communist state, and nothing about the climate problem makes me think otherwise. I favor regulating our current capitalist system

67 Mitz-Woo, In preparation.
68 X, 1964.

and holding corporations responsible. By my judgment, this promises the best chance of success.

Andrew: I'm sorry to break it to you, but your approach seems centrist to a fault. We aren't going to be able to respond responsibly to climate change without radically changing our ways of thinking and living. We can't solve the problem until we become an anti-racist, anti-sexist, anti-capitalist society.[69] Environmental racism allows white people to shield themselves from the consequences of their pollution by dumping them on people that don't look like them. Our attitudes of domination towards the environment reflect the same attitudes of domination we have towards women. Capitalism incentivizes endless production, consumption, and exploitation, which all need to go down.

Eliza: (*stands and begins pacing*) Centrism is a bad word in some circles, but I think it's what's needed. Climate change is nonpartisan, or at least should be. It threatens most central values: equity, justice, community, but also individual freedom, property rights, national identity ... lots of things. We need folks to come together on this, conservatives as well as progressives. That's what I'm fighting for.

Hope: I also want to make this nonpartisan, but lifestyles and economic systems *will* have to change, Eliza.

Eliza: Perhaps. I still have some hope for a green energy revolution; so, maybe lifestyles will have to change, but not as dramatically as Andrew would like.

Andrew: I think you underestimate the scope of the problem.

Awkward silence.

Eliza: Anyway, my larger point was that, whatever climate action we want to see, the government is who we need to compel. If a bridge needs repairing, it is the government's

---

69   See Dialogue 4 for further discussion of this point.

role to fix it. If they don't, it's a problem, but it's not like you can blame me for the problem.[70] It's not up to me to go out trying to fix the bridge myself. Telling me to do so, and moralizing if I don't, would be silly and counterproductive. This is what I was getting at when I suggested we shouldn't suddenly embrace a strange form of anarchism when it comes to solving the climate problem. We don't elsewhere.

Andrew: Look, maybe we could leave climate action to the government *if* we had an effective government. But it doesn't look as though we do on this issue, and that's a nonpartisan comment. We've been struggling for thirty years, since the early '90s, across Democratic and Republican administrations. Lots of other countries are struggling too.

Eliza: I know! That's why I think my main responsibility is to *get* the government to act responsibly. In the first instance, I think everyone should back candidates who care about climate change and will fight for aggressive climate action. We shouldn't vote for politicians who won't.[71] Political action should be the priority: not personal lifestyle change.

Hope: I am inclined to agree with you, but you worried earlier that people shouldn't have to offset their carbon *specifically*, because there are a bunch of other injustices they also contribute to. Couldn't a similar argument be made here? What about the other political issues? You know, the economy, national defense, even moral issues, like abortion. Why shouldn't voters focus on these issues, rather than climate change?

70   Sinnott-Armstrong, 2005, p. 295.
71   Note that Eliza does not see voting as the limit of individual political responsibility (e.g., later in this chapter, she argues for global institutional change). For a useful illustration of how such responsibility may play out when delegation to existing institutions fails, see Gardiner's "get armed and go West" example of breakdowns in security (Gardiner 2017d, pp. 38–39).

Eliza: I guess I do think climate change is more important. It threatens to overwhelm all the others and soon. I see voting for climate deniers as like voting for appeasement or isolationism in the run up to World War 2. You can see why people might have done it. Why they would want to avoid war and its costs; why they might have had other priorities. But in the longer run, it was a failure, since what happened then swamped everything else for a while, and then it took a long time to recover. If we can get the climate action we need and quickly, then we can get back to prioritizing other, more normal issues.

Andrew: But our individual responsibility isn't just to vote! It is also to push back against climate misinformation, stop climate appeasement, take on the fossil fuel lobby, dismantle capitalism, *and* reduce our personal emissions. All of the above.

Eliza: I'm still not sure that *all* of that is my responsibility, especially when I'm already dedicating a large part of my life to fighting for climate justice.

Andrew: (*chuckles*) Full circle, huh? It sounds like we're right back to the question of integrity.

What Governmental Change Should Individuals Be Fighting for?

Hope: Well, it's already getting late, and we haven't written anything down. (*smiling*) So, let's not repeat that discussion. I want to ask Eliza a question, though.

Eliza: Uh oh. (*pretends to brace herself*) Shoot.

Hope: You think individuals should primarily be fighting for collective change, not worrying about their individual emissions so much. So, what collective change should individuals be fighting for?

Eliza: Well, structural overhaul and decarbonization, of course.

Hope: So, individuals should vote for whatever candidate is most likely to do that?

Eliza: Yeah. At least that.

Andrew: We need to ensure the transition away from fossil fuels is a just one. This means that wealthy countries take the lead on decarbonization, that marginalized and vulnerable communities are protected, that change is made in ways that are anti-racist and anti-sexist.

Hope: (*scribbling*) So, these seem to be principles we all agree on, and can use.

Eliza: I think we need to go even deeper. I have some pretty radical ideas.

Andrew: My favorite kind! Pray tell.

Eliza: Right, here's what I think. We each have a moral responsibility to others, and not only to people alive today but to future people and the rest of nature. However, we can't discharge those responsibilities by ourselves. For example, we can't protect the global climate system alone; that requires the cooperation of more or less all of humanity. Given that, our obligations become collective, to cooperate together to put in place just institutions. If we fail to do that, then we've failed to discharge our responsibilities, and that's something for which we're liable to criticism.

Hope: Nothing too radical yet.

Eliza: Just you wait. Here's my proposal: because climate change is an intergenerational problem, we need to create new institutions that can protect the interests of the future.

Andrew: You mean like by having an ombudsman for future generations, or perhaps special votes for the young, or a third chamber of Congress, an assembly for future generations?

Eliza: Yes. (*unenthusiastically*) Those are some of the ideas.

Hope: That doesn't sound like a ringing endorsement! Do you have something else in mind?

Eliza: Oh, I don't mean to be dismissive. I think any of those things might help to some extent. But I still think they don't go far enough.

Hope: What do you mean? They all sounds pretty radical to me. What more do you want?

Eliza: Well, perhaps I'm more radical than you think. *(winks)* Look, I worry that this is fundamentally a global and intergenerational problem. But those solutions are at the national level. In my view, the ultimate objective needs to be a global institutional body, and national solutions don't establish one. To be honest, you were the one to inspire this idea, Hope!

Hope: *(surprised)* Me?

Eliza: Yeah. After you explained that current generations might not consider the interests of future people, I started thinking about how we might fix that problem.

Hope: That's awesome. What's your proposal?

Eliza: I think we need a global constitutional convention.[72] Something in the spirit of the American constitutional convention of 1787, or the numerous other constitutional conventions that have taken place around the world since then. Ours would be a deliberative body tasked with establishing a system of governance for dealing with intergenerational issues. It would directly confront the global nature of the problem. It would also coordinate a response. We would get beyond the patchwork solutions generated by a bottom-up approach: you know, an ombudsman here, youth votes there, a third chamber over there. It would also look at alternatives and do a comparative assessment. It should try to figure out what the best overall global system would look like, how to make sure that it gives appropriate – but not excessive – respect to other important institutions, and how to protect individuals and communities from abuses of power. ... *(looking to Hope)* The global constitutional convention would, of course, have to be

---

72 Gardiner, 2014; Gardiner, 2019a.

carefully designed, with issues of justice and intergenerational responsibility very much in mind.

Andrew: I love it!

Hope: (*laughing*) That does sound radical – I'm intrigued. But now I think we're way off topic. We need to create this list.

Andrew: Well, perhaps we can start by proposing something like an institution for the future here at the university! It would not be global, but we could recommend the university establish a body responsible for representing the interests of future students.

Eliza: Great!

Hope: That's a wonderful idea. Do we also agree that we should invite students to reflect on how to reduce their carbon emissions? Could we offer some specific ideas there, so long as we take care not to make them elitist or insensitive to people's different situations?

Eliza: I can accept that. But can we make clear that collective action is also essential? And offer some ideas there?

Hope: How about we begin by creating a group to make some *concrete* demands of the university? Demand aggressive mitigation goals …

Andrew: (*interrupting excitedly*) Like, electrify all university vehicles in ten years! Make access to public transport free for students, faculty and staff! The whole university carbon-neutral by 2035!

Eliza: We should also encourage other universities to establish similar goals, and pledge to hold each other accountable.[73] And let's explore what universities might do to galvanize city-wide or state-wide change.

Andrew: I'd like us to encourage folks to lift up ongoing movements like Fridays for Future or Sunrise, or to start their own. Protest business-as-usual with all the energy we can summon! Demand equitable adaptation, divestment

---

73 For example, see Second Nature's Climate Commitments for colleges and universities: https://secondnature.org/signatory-handbook/the-commitments/.

from fossil fuels, and steps towards decarbonizing infrastructure. Sit-in! Teach-in! March!

Eliza: (*smiles*) Collective action is so full of possibilities.

Andrew: Individual action too!

Hope: Yes, it's easy for this conversation to get too polarized and abstract. When we start getting more concrete and looking forward, it becomes much clearer that there is a lot to get done, and all of us have a role to play.

Andrew: (*grinning*) So, we're agreed! (*holds up a pen dramatically*) Let's make a difference![74]

---

74 For an excellent survey article on the ethics of individual emissions, see Fragnière, 2016.

Dialogue 4   International Justice

Sometime in the 2020s[1]

Part 1: Mitigation and Burden Sharing

Hope fiddles with her hands as the others trickle into the room, excitement bubbling over into nervous energy. In truth, the regular negotiations were not going well. While attending a rather boring official session the previous day, she had the idea of arranging this alternative "side-meeting" of junior negotiators from a diversity of nations. Initially, she had worried that the others wouldn't be willing to take the time out of their already busy schedules, but this anxiety proved groundless. Almost everyone accepted. It was happening.

Informal as this meeting would be, she can't shake the feeling that it might be something she remembers the rest of her life. Maybe this is it. If this group can reach a consensus regarding the thorny issues of international responsibility and climate equity, they can take it to their bosses.

At that moment, Richard, a fellow American representative, walks over.

Richard: Hi Hope! Thanks for organizing this.
Hope: Of course! I was just about to get started. (*clears throat*) Hi everyone! Thanks for coming. Are we all ready to have a new go at this? (*Murmurs of assent*) I'm Hope,

---

1 See also chapter 5.

DOI: 10.4324/9781003123408-4

representing the United States. This is Richard, a member of our new climate caucus in the House. Despite our many political differences ...

Hope smiles; Richard smiles back.

... we are both committed to pushing climate action forward quickly. I trust all of us here share that goal.

Richard: Thanks, Hope. I'm here because Hope and I are seeking an approach that has a chance of getting a strong consensus in both Congress and the country to get us past the current mess. We need to set the United States on a path towards substantive climate action that will survive changes in administrations.

Susan: I'm Susan, from the United Kingdom. We believe that the international community must also reassess where we are, and recalibrate our collective response. Together, we need to agree on a way forward that all of us can support, away from the media glare. (*more murmurs of agreement*) We all accept the goals of avoiding a 2°C temperature rise and aspiring to get closer to 1.5. For a while now, and especially under the Paris agreement, countries have been setting their own mitigation goals, according to their views of what they can reasonably achieve. This approach agreement is basically voluntary, with no meaningful enforcement mechanisms. Unfortunately, it continues to fall short. I believe that we all recognize this, privately if not publicly. So, my main purpose today is to see if we can quickly ramp up our efforts, by coming to a wider agreement on what it is reasonable for us to expect of each other, and how to deliver that.

Amina: Hello. Amina, from the Maldives. My government agrees. Deeper progress is needed. We also recognize that we've been down this road before, many times. There have been countless efforts in the past, but they've all come up short – usually after being undermined by certain big countries, and often after they themselves have played a big role in designing the agreements.

Richard: (*in a whisper, to Hope*) She means us, right?

Hope: (*grimacing*) Well, yes ... but not *only* us.

Amina: So, before we consider any details, I'd like to see if there's a commitment to the basic norms established long ago: to equity and "common but differentiated" responsibilities. Without that agreement, this initiative is likely to fail, like all the others before.

Hope: We are committed to treating other countries fairly, so long as we are treated fairly. We also accept that there may be differences in responsibilities, though we'd expect that all countries should accept meaningful responsibilities, and that those differences that remain should be equitable.

A general rumble of approval.

Why Justice?

Richard: Wait. I don't see why we need to talk about equity and responsibilities at all. Isn't it pretty straightforward? We can just put a reasonable price on carbon, ratchet it up over time, and then let the market determine who does what. When we correct for market failure by making people face the real costs of carbon pollution, they will make different decisions. Folks will shift their consumption away from fossil fuels, and toward more environmentally responsible products, services, and activities. Companies will do the same. The invisible hand, aided by a sensible intervention to correct for market failure. What's fairness and equity got to do with it?

Indira: I'm Indira, representing India. I'll say something to that. We worry that the rich will just buy their way out, and the poor will end up shouldering most of the burden. If the rich want to continue to consume more fossil fuels, they will simply pay the higher prices. But the poor, and perhaps even the average person, might not have that option, at least without sacrificing important things, like food, heating in the winter, getting to work. Do you remember what happened years ago with biofuels? When the United States changed its ethanol rules and removed nearly five billion bushels of corn from the market? Some argue that this triggered spikes in food prices, causing civil unrest worldwide, including the Arab Spring.[2]

Richard: I share your concern. My constituents would also find that scenario hard to accept. We can't have rich, city folks swanning around in gas-guzzling luxury cars while we're freezing to death out in the country. For this to work, there has to be some reasonable sense of burden-sharing. We need to feel that we're all in this together, carrying proportional burdens.

Indira: So, you agree that fairness in burden-sharing matters?

Richard: Yes.

Amina: Do you also accept the general ethical consensus on this?[3] That the richer, more developed countries should take on more substantial burdens in the transition away from fossil fuels, at least in the short to medium term?[4] The commitment to equity and "common but differentiated" responsibilities?

Richard: (*looking concerned*) Maybe. It might depend on what you mean by "differentiated."

Indira: What would you propose?

2   Bar-Yam, *et al.*, 2015.

3   Gardiner 2004, p. 579; Gardiner 2016b, p. 100. See also Shue 1999, Singer, 2016, and Morrow 2017.

4   The distinction between "developed" and "developing" countries (and who falls on which side of that division) is controversial in itself and may have "outlived its usefulness" (e.g., Olopade 2014; Khokhar, 2015). We continue to employ it here in part because the United Nations continues to do so in talking about climate (e.g., in the Paris agreement), and because of its role in the wider literature.

Richard: Suppose we start from where people already are. We know that we should be aiming at net zero by around 2050. Can't we just say that each country should eliminate its emissions by then?

Setting a Carbon Budget

Susan: I want to be sure we're all on the same page. When scientists say "net-zero by 2050," they do not mean "do what you like until 2049 and then cut to zero." A better way to talk is in terms of a carbon budget: the total amount of carbon dioxide we can emit if we are to have a good shot at meeting the 2°C or 1.5°C targets.

Richard: Remind me: how much is that?

Susan: As a rough guide, by the end of 2019, the total carbon budget for 1.5 degrees had been reduced by around 2390 gigatonnes of carbon dioxide ($GtCO_2$). If we wanted to aim at a 67% probability of limiting warming to 1.5°C, countries could only emit another 400 $GtCO_2$ total from that point.[5]

Richard: And how soon will we use it up?

Susan: In 2019, global emissions were depleting the budget by around 42 $GtCO_2$ each year. So, at the rate we're going, we'll have used up the whole budget for 1.5°C sometime in 2028.

Richard: It is rough; I know.

Indira: Things would have been a lot better if we'd started much earlier, as many countries were arguing thirty years ago. There have been several lost decades.

Amina: We must also take seriously that our situation may be even worse than Susan indicated. It is far from clear that the 2019 numbers represent a peak. Projections in 2019, prior to the pandemic, suggested that annual global emissions might be up to 60 $GtCO_2$ per year by 2030, or 54 $GtCO_2$ if the 2015 Paris commitments were

5 IPCC, 2021b, p. 38.

fully implemented.[6] The coronavirus only depressed global emissions for a year or so.[7] In other words, what's required is an enormous turnaround.

Susan: We'd have a bigger budget remaining if we were comfortable with only a 50% chance of not busting 1.5°C. That would leave us about 500 $GtCO_2$ left to emit.[8] With this carbon budget, we'd have fifteen years at current emission levels from 2019, so until roughly 2031. But, of course, then we'd have to be at zero *instantly*, and for the long term.

Amina: We are uncomfortable with only a 50% chance of holding to 1.5°C. It seems like a betrayal. More than 1.5°C spells disaster for my island and many populations worldwide.

Richard: Okay. How do we improve our odds of holding warming to 1.5°C?

Susan: If we are comfortable with a 66% chance, global emissions should be cut in half by 2030, and get to net-zero around 2050.

Richard: That's grim. How aggressively would we need to cut to achieve the 2°C target?

Susan: For a 67% chance, global emissions have to come down 25% by 2030, and to net-zero by 2070.[9]

Richard: Okay. Suppose we aim for 1.5°C. Starting where we are, couldn't every country agree to cut, say, 4.5% of their own emissions each year until we reach our reduction goals? Or if we fallback to the 2-degree target, we could all agree to a 2.5% cut each year? Either way, everyone would have to do the same thing, relatively speaking,

---

6 UNEP, 2019, p. XVIII.
7 McSweeny & Tandon, 2020.
8 IPCC, 2021b, p. 38.
9 IPCC, 2018, p. 12. There are some uncertainties. For example, carbon released from melting permafrost and methane from wetlands could reduce the carbon budget by up to 100 $GtCO_2$. In addition, assumptions about mitigating other things apart from CO_2 could alter the budget by 220 $GtCO_2$ in either direction.

since it is an equal percentage cut for everyone. But the lower emitters would have to do less in absolute terms, so there is still a sense of differentiated burdens. Is that fair enough for you?

Indira: Not really.

Richard: *(surprised)* Why not?

Polluter Pays Principle

Marcia: Hi, I'm Marcia, representing Brazil. I'd like to answer your question. One reason is causal responsibility. I think an equal percentage cut lets big emitters off the hook. If a company pumps waste into a river, shouldn't it be their responsibility to clean up?[10] Equally distributing this burden across everyone would be unfair. You broke it, you fix it!

Richard: In principle, I agree. The company made the mess; they should clean it up.

Marcia: So, some countries, like the United States, have historically emitted more carbon than other nations, and *far* more than most developing nations. Since you are the biggest causes of the problem, shouldn't you also be responsible for doing most to clean up your own mess?[11] You know, "polluter pays": the idea that whoever is responsible for damage to the environment should bear the costs associated with that damage.[12] It's a well-established principle of environmental law.[13] It ensures

10   Famously, Brazil emphasized the relevance of historical emissions to burden sharing in 1997 during the negotiations leading up to the Kyoto Protocol; thus, it came to be known as "the Brazilian Proposal" (den Elzen, 1999).

11   Caney, 2010.

12   Cordato, 2001, p. i.

13   In international climate change law, the polluter pays principle was formally established by the Rio Declaration, Principle 16. The principle is also widely accepted in general environmental policy (European Court of Auditors, 2021; OECD, 2021).

that entities are accountable for the consequences of their actions and incentivizes them to be environmentally conscious.

Richard: Look, I agree with the polluter pays principle most of the time, but I don't think it's fair in this case.

Marcia: I'm pleased to hear that you are so committed to fairness. But why do you think my proposal is not fair?

Richard: Well, for most of the two hundred years since the Industrial Revolution, people were blissfully ignorant of the fact that emissions caused a greenhouse effect. It's a relatively recent phenomenon.[14]

Amina: I disagree. The idea of a human-caused greenhouse effect has been around for more than a century.[15] In 1965, scientists briefed your President Johnson about it, warning that we were increasing the atmospheric concentration and that a doubling of CO_2 would lead to a global temperature rise of 1.8–12.0°.[16] The President even gave a speech warning that we were conducting "a vast geophysical experiment," and emphasizing that "the longer

14 The wording here is drawn from a remark made by Todd Stern, U.S. climate envoy under the Obama administration. See Revkin & Zeller, 2009.

15 There were important early contributions from Joseph Fourier (1824) and Claude Pouille (1836). In 1856, Eunice Newton Foote, a female scientist and campaigner for women's rights, demonstrated the heating effects of carbon dioxide and water vapor, foreshadowing John Tyndall's more famous experiments published three years later (Huddleston 2019). In 1896, Svante Arrhenius, a Swedish chemist, suggested that burning coal could warm the planet (UCAR, 2021).

16 The report states: "By the year 2000 the increase in atmospheric CO_2 will be close to 25%. This may be sufficient to produce measurable and perhaps marked changes in climate ... With a 25% increase in atmospheric CO_2, the average temperature near the earth's surface could increase from 0.6 to 4.0 degrees C, depending on the behavior of the atmospheric water vapor content ... *A doubling of CO_2 in the air, which would happen if a little more than half of the fossil fuel reserves were consumed, would have about three times the effect of a twenty-five percent increase [i.e., 1.8–12.0°C]*" (Environmental Pollution Panel and President's Science Advisory Committee 1965, pp. 121–127).

we wait to act, the greater the dangers and the larger the problem."[17]

Richard: Hmm. Still, our leaders get lots of warnings. They can't act on all of them. As I understand things, it wasn't until the 1990s that there was a strong consensus on the rest of the science. You know – that it became clear there were no other explanations for the rise in GHG concentrations and no negative feedbacks that would offset them.

Amina: I'm not sure previous uncertainty is a good enough excuse for inaction, considering the threat. What would you think about a business that was warned that its pollution was likely to harm people but continued to dump its waste into the river? What would you say if the company *dramatically increased* the amount it was dumping *after the warning*? Don't you think they would be *negligent*? I do. The company should have taken reasonable precautions, even as they were waiting for the full facts to emerge. They are liable for not doing so. You're the same. Arguably, your country knew *enough* in 1965 not to go on a fossil fuel *binge* for another fifty years.

Richard: I think you're being too judgmental.

Amina: How convenient.

Appropriator Pays Principle

Indira: There's another issue.

Richard: Pray tell.

Indira: In using up so much of the carbon budget, you effectively deprived us of the opportunity to develop using fossil fuels. You took more than your fair share of that

---

17 See "Climatic Effects of Pollution" in Environmental Pollution Panel & President's Science Advisory Committee, 1965.

valuable resource, and in doing so you appropriated *our share*; you should compensate us for that.

Susan: I'm not sure I follow. Isn't that the same as polluter pays? What difference does talk of "appropriating shares" make? Isn't it the costs imposed by using up the carbon budget that matter?

Amina: I'd say that the issues are similar, but not the same. Recall that the polluter pays principle says that whoever is responsible for *damage* to the environment should bear the costs associated with that damage. The *appropriator pays principle* says something a little different: that those responsible for *depriving others of their fair share* of a resource should compensate them for that deprivation.

Indira: Ah. So, one difference is that the appropriator pays principle does not assume that the problem is *damage,* and *to the environment* as such. The key issue is someone's being *deprived,* and *of their fair share.*

Amina: Another difference is that the appropriator pays principle specifies more precisely what should be done. It says that those deprived should be compensated by the appropriator. The polluter pays principle states only that the polluter should bear the costs – it leaves the rest open. For example, the government might take the money for itself, without really compensating anyone. That often happens. Furthermore, it might "fix" only whatever part of the problem bothers it. I could go on ....

Richard: Don't bother. I see now that they are different principles; but I still think both fall to the same objection. We *didn't know* fossil fuel use would need to be stopped! We had no idea it would turn out to be a scarce resource.

Indira: We disagree about that. But my point is different. Even if you were blamelessly ignorant that you were using up the carbon budget, I don't think that gets you off the hook. We should still get some compensation. Think of an analogy. Suppose you took my slices of pizza out of the fridge, thinking they were leftovers that no one wanted. You might be blamelessly ignorant of the fact that they were mine. Still, I think you'd owe me some compensation. After all, I suddenly don't have any lunch.[18] Surely, it's not fair for that to be simply *my* problem – you brought it about! Similarly, in the climate case, we think you should assist us in developing cleanly, to make up for depriving us of the fossil fuel path.

Richard: I still don't agree. But there's another objection to all this liability talk that you are *conveniently* ignoring. The Industrial Revolution started in the eighteenth century. It's not fair to hold my generation – indeed living Americans in general – responsible for something that started hundreds of years ago. Most of the emitters are dead. You're not making *the polluters* pay – what you're actually trying to do is to make other people, their descendants, pay instead.[19] But how is it fair to hold the American people liable for things that many of them never did? It's not our fault who our ancestors are, and we shouldn't be held liable for the sin – or debts – of our forebears. In addition, for many Americans, our families have only been here for a generation or two anyway. So, it is not even *our* ancestors!

Hope: That's a bit misleading, Richard. Many of our emissions have occurred in the last fifty years. Indeed, more than half of global emissions have occurred since the late eighties, when scientists were busy sounding the alarm.[20] So, to a significant degree, we're *not* talking about the

---

18 Gardiner & Weisbach, 2016, pp. 114–115.
19 Caney, 2010a, pp. 127–130.
20 Frumhoff, 2014.

emissions of dead people.[21] A big chunk of emissions have been caused by folks still around now.

Amina: I have a more general worry. I don't like the whole "dead emitters" attitude. You're trying to have it both ways.[22]

Richard: What do you mean?

Amina: On the one hand, think about benefits. You're not responsible for creating most of the assets and infrastructure of the United States, right? You inherited them from past generations. You had no choice in that; it was "beyond your control." Yet, you still think *you're entitled to keep those assets*. In fact, you'd insist on it. You don't complain about unfairness there! But, on the other hand, when it comes to the costs and liabilities you inherit – the burdens inflicted by your predecessors in building up those assets – you want to wash your hands of them. That's why I say you're trying to have it both ways.

Marcia: Reminds me of what used to be called "lemon socialism": privatize the benefits; socialize the costs.[23]

Amina: Given the current international system, I'd prefer to think of it as "lemon capitalism"! Privatizing benefits is a very obvious feature of global capitalism; quietly socializing the costs is less so.

Beneficiary Pays Principle

Indira: This discussion gives me an idea. Maybe we don't need to get hung up on who knew what and when, and who was alive when. Whatever else one might say, current Americans, Europeans, Russians, and the Chinese have clearly *benefited* from their countries burning fossil fuels, now and in the past. Given that, we think that it is only fair that they should contribute more to solving

21 Meyer, 2013, pp. 605–606.
22 Gardiner, 2011a, pp. 418–419.
23 Shue, 2015, p. 20. This term was coined by Mark J. Green (Green, 1974).

the problem. Beneficiaries should pay. We should adopt a *beneficiary pays principle*.

Marcia: I like that idea. It seems much better than letting the burdens fall on the victims of climate change. After all, why should the victims pay? Surely there's a *stronger* claim that beneficiaries should pay than victims who have contributed very little to the problem.

Richard: But it is not our fault that we've benefitted! We had no choice in the matter. You shouldn't hold us accountable for something that we had no control over!

Indira: That may be true, but then we're right back to Amina's point about lemon capitalism. If a nation's citizens acquire burdens only when they themselves are responsible for bringing them about, the same should go for benefits. Otherwise, you seem to be worried about causality, choice, and control only when it benefits you to care about them. That's hard to take seriously.

Amina: Exactly. All we require is consistency. I think I can speak for all developing countries when I say that we would be happy to drop the liability angle in exchange for one other thing.

Richard: Really? What's that? (*with a tone of surprise*)

Amina: I propose that we all agree that current citizens of major historical emitters, like the United States, are not responsible for the costs of their countries' past behavior, *so long as* they all agree that they are also not entitled to the benefits they've inherited. In other words, we agree to share the costs of industrialization and development globally, but only on the condition that we all share in the benefits too.[24]

Richard: Preposterous. You are not getting a slice of our economy or our civilization. We are not giving you Manhattan or Kansas!

24 Gardiner, 2011a, pp. 418–419; Gardiner & Weisbach, 2016, p. 118.

Amina: Don't be silly. We're not asking for Manhattan or Kansas.

Richard: What are you asking for?

Indira: Just give us our share of the resources. How about this? Something like a global resource dividend would do it. Everyone on the planet gets a stake in the returns from natural resource transactions. That does not look especially onerous or disruptive.[25]

Richard looks at Hope in exasperation. She smiles back weakly. Honestly, it didn't sound that unfair to her.

Indira: We might even go one better. Suppose we use the resources to transition to clean energy. Build solar power plants, instead of coal-fired ones. It is a win–win on multiple fronts. You escape a liability issue, we'll get to develop quickly and cleanly, and we'll all avoid dangerous climate change!

Richard: How much money are you talking about? The United States won't finance an entire overhaul of India's infrastructure. That's never going to fly.

Hope: I wonder if we might converge on a more limited issue. We disagree about the relevance of emissions before the early 1990s. But that's more than thirty years ago now. The United States has been attempting to address climate change ever since, including by making agreements in Rio, Kyoto, and Paris. But we've largely failed. None of that is on the back of people in the 1800s. So, we might be open to taking responsibility for more recent emissions.

Richard: I have to disagree with my colleague here. What's done is done; politically, we've got to be forward-looking.

Carlos: I'd like to get in if I may. My name is Carlos. I have not spoken yet. I represent Bolivia.

Hope: You're welcome.

25 Beitz, 1999, pp. 136–142; Pogge, 2002, ch. 8.

Carlos: I'd suggest reconsidering. Politically, we must address the issues people care about. Many countries care about the United States' causal responsibility and its past failures. There needs to be a general framework of trust if we are to make progress. Saying you refuse to take responsibility for your past mistakes tends to undermine that.

Richard: (*takes a breath*) Okay. Suppose I compromise and accept more recent historical responsibility. I'm not sure that I can get the rest of my party on board with that, but I see where you are coming from.

Indira: We appreciate that.

Richard: Nevertheless, I'm not sure whether it makes a big difference. Notice that we're not the worst offenders. Our emissions have been coming down since 1990, while they've been exploding in other places. Look at China. It's now the biggest emitter, and accounts for more than a quarter of global emissions – up from about 4 $GtCO_2$ per year in 1990 to more than 13 per year now.[26] Your country, India, has been rapidly increasing its emissions, at a rate of about 3.7% per year. By contrast, the United States has been slowly reducing its emissions by about 0.1% across the decade, and overall has remained at 6–7 $GtCO_2$ per year since the early '90s. We also have plans to decarbonize. So, I'm not sure that it is fair to keep singling us out. Unless China and India get their emissions under control, we'll all be toast anyway. Why let them off the hook?

Kim: I represent China, and I resent the deflection. You ignore our population and per-capita numbers. We have many more people than you. The United States emits more than twice as much per person as China.

26 UNEP, 2019, pp. 5-6.

Hope: (*quickly*) I think all of us in this room agree that no
            country can continue on a fossil fuel-intensive develop-
            ment path. That would completely blow any reasonable
            carbon budget. So, we should not let *any* countries off
            the hook. But we can admit, Richard, that there are dif-
            ferences in circumstances between our emissions and
            these other countries. For example, China as a whole
            used to be poor, although it is not anymore; and India is
            still developing.

Kim and Indira nod.

Richard: I don't think we can keep making excuses for other
            countries. The situation is too desperate.
Amina: We agree!

*Richard winces, registering the irony. The room has grown tense.
Hope realizes an intervention is needed.*

Ability to Pay Principle

Hope: (*hastily*) How about coming at this from a different angle?
            Suppose we set aside history for a moment. (*Indira's eyes
            harden*) Instead, we simply ask who has the most *capac-
            ity* to address the problem, according to an ability to pay
            principle. It seems the rich countries, like my own, have a
            far greater ability to bear the burdens of a transition than
            many other countries, especially developing countries.
            We have more resources, more developed infrastructure,
            more technological know-how, and so on. Surely, we
            should contribute more for that reason alone?
Amina: Maybe. An ability to pay principle would suggest that
            the United States, the Europeans, and China should con-
            tribute more, and without getting bogged down in other
            issues, like liability. If the results are broadly the same,
            perhaps we don't have to prosecute the details of which

particular approach to fairness is the best? We could just focus on the least complicated.

Hope: That's what I was thinking.

Susan: Hold on. What about those who have invested quite a lot in clean development and decarbonization over the last few decades? Surely, we should not be on the hook to the same extent as those who have continued on a high emissions pathway? Isn't that unfair?

Hope: Would you be open to some combination of ability to pay and recent historical responsibility?

Susan: A hybrid approach? That might work. We'd have to think through the details.

Amina: How about a 50–50 split between capacity and historical responsibility?[27]

Indira: We might favor something more like 80% historical responsibility, and 20% current capacity.

Marcia: Pulling percentages out of the air all sounds a bit arbitrary to me. Still, the approach might work, if we can find a rough consensus everyone could live with ....

Richard: Wait. I think there's a deeper problem. We have the ability primarily because we have the resources. And we're entitled to those resources!

Equal per Capita Principle

Hope: That reminds me of a different approach. Perhaps we should talk about what *entitlements* people might have to the remaining carbon budget?

Richard winces. Hope can tell he's annoyed by her choice of words.

Amina: Agreed. We believe that a natural default position is to say that every person alive today should get an equal share.[28]

27 Baer, 2010, p. 224.
28 Agarwal & Narain, 1991, pp. 9–12; Jamieson, 2001; Singer, 2016, pp. 41–46.

Richard: Why is this "a natural default position"?

Amina: One reason would be that all human beings are equally important from the moral point of view, so that each person should be treated with equal concern and respect. This implies that all have equal rights in general, making it difficult to justify departures from equality, especially when we're talking about resources that are essential to well-being. So, it seems to us that, other things being equal, no one deserves less than an equal share of the carbon budget.[29] There are also precedents for treating common pool resources this way, especially when they exist outside of the control of individuals or states.[30]

Marcia: Another reason might be that it is just easier to start there. It is a simple proposal, and one that still addresses the sharp inequalities of the status quo. It also seems to gel with other approaches.[31] Beginning with a strong presumption in favor of equality shows basic respect to everyone's interests, which seems likely to reduce conflict. In this case, it also seems to promote well-being in the long term, since everyone is taken into account, even the poor who currently emit little and whose interests may need special protection.[32] So, the equal-per-capita approach at least seems to move the discussion in the right direction.

Indira: We may be able to live with that. We still believe historical emissions should matter. Nevertheless, it's clear that the developed countries need to take on a greater burden than developing countries under the equal per capita proposal as well. After all, their per capita emissions are far higher, and that's unjust even if we ignore history.

29   Baer, 2002, p. 401.
30   Baer, *et al.*, 2000.
31   Singer, 2016, pp. 54–55.
32   Singer, 2016, pp. 54–55..

Richard: (*looking thoughtful*) How would this idea play out? Could you remind me of various countries' per capita emissions? I know that the United States is at about 15 tonnes.

Susan: In 2016, the global average was around 5 tonnes of carbon dioxide per person.[33] As you say, the United States was at about 15, as were Australia and Canada. Russia was at 12, China at 7.2, the European Union at 6.5, Brazil at 2, Bangladesh at 0.5, Haiti at 0.3, and Ethiopia at 0.1. My country – the UK – was just over 5.

Richard: That's a pretty big disparity. What is the per capita target?

Marcia: Well, that depends on the carbon budget, as you know. But if we're aiming for 1.5°C, 2–2.5 tonnes a year per person for 2030 is pretty plausible.

Richard: Whoa! The United States isn't going to be able to get down that low in ten years. That would be something like an 85% cut for us! The current aim of 50% is already a big challenge. I fear that 85% is unworkable.

Indira: Perhaps we could deal with this if we allow countries using less than their entitlements to sell that to those over budget? That's always been a key part of most versions of the equal-per-capita approach. Countries like the United States, who would otherwise struggle to make cuts quickly enough, could buy permits to emit from the low emitters, who could then use those funds to fund clean energy transition, or reduce poverty, or for other development goals.

Hope: One advantage of this approach is that, since those who benefit – those selling permits – are basically the less developed countries, we preserve the spirit of "common, but differentiated responsibilities," and we get compensation to low emitters. Notably, we get similar results to under principles of historical responsibility, but by different means.

Amina: But your burdens would still be *less* than if we factored in historical responsibility. Right?

Hope: True; but it's worth realizing that developed countries would still need to bear a larger burden and contribute more to solving the problem than less developed countries. However, this is because of their *current* unsustainability, so the approach avoids the controversy about past emissions. I think the United States might be able to get on board with that. Do you agree, Richard?

Richard shrugs noncommittedly.

Indira: This discussion is worthwhile. It's very important to realize how, even ignoring history, the developed countries should take on a greater burden.

Hope: I sense a "but" coming ...

Anti-Poverty Principles

Indira: How perceptive of you! We feel that the equal per capita approach ignores other central issues.

Hope: Such as?

Indira: We're doubtful that the basic equality argument works. People don't care about carbon emissions for their own sake. Emissions are not inherently valuable, like well-being or dignity. Instead, their value is *merely instrumental*, as a means to other ends. In other words, emissions are important only insofar as they affect what really matters, which is the impacts on people and especially their well-being. Paying too much attention to emissions – all those very precise numbers! – makes a fetish out of them.

Richard: I agree!

Indira: Moreover, what seems to matter for well-being is *access to energy*. Some countries have very low per capita emissions already – lower than even very stringent targets. But, typically, our emissions are low *because* our infrastructures are so underdeveloped, which is *because* of our poverty. Given that, we are concerned that any obligation to *keep* our per capita emissions very low will severely constrain our ability to combat poverty and deliver a reasonable standard of living for our citizens. This is one of the main reasons why we think any sensible proposal should factor in historical responsibility. If the approach is simply equal-per-capita, it risks perpetuating massive global economic inequality going forward.

Richard: Hold on. It sounds like you might be arguing that you should be exempt from limits. We obviously can't have that. If India and countries like it continue on a fossil fuel development path, then any reasonable carbon budget will be completely blown. The same is true for China, by the way.

Kim, the Chinese representative, tenses, but says nothing.

Indira: We are not insisting on a fossil fuel path. We are saying that because many of our citizens are still very poor, we deserve additional outside assistance to develop sustainably. The richer countries should help us to take a clean development path.

Amina: We agree. Protecting the least well-off is at the center of many approaches to fairness. So, if you think that people's interests matter more when they are badly off, then we have another reason to think that richer countries should take more of the burden.

Indira: This brings us to an important point, one that the equal per capita approach ignores. Many of the world's current emissions are for luxuries, others are necessary for subsistence. Can we all agree that luxury emissions ought to be cut first? Even in an emergency, you don't ask some to give up their blankets so that others can keep their jewelry.[34]

Amina: We agree. Subsistence emissions must be taken incredibly seriously. For some, there are no real alternatives to using fossil fuels to fulfill their basic needs, like for food, heating, making a living. It can't be fair to deprive them of the fundamentals of a decent life.

Peter: Can I get in? I'm Peter. I'm an activist, representing the Anti-Poverty League, an NGO.[35] I'm thrilled to hear the focus on protecting subsistence emissions. It seems that a lot of this comes down to protecting the most vulnerable, especially the global poor. That should be our priority. Poverty is the fundamental underlying issue. I'm talking about the millions living in truly desperate circumstances, on a dollar or two a day. They face severe malnutrition, squalid conditions, and circumstances that are barely imaginable to most affluent Westerners.

Richard: I believe I can speak for the United States that caring for the worst off – including the very poor – is at the center of our ideals. We see ourselves as a land of opportunity where people can escape poverty, and anyone can find happiness if they are prepared to work for it.

Peter: That's wonderful to hear. My organization believes that poverty is such a scourge of humanity that we shouldn't do *anything* that increases it or delays its eradication. In fact, my organization proposes that we should see poverty eradication as the fundamental international objective for climate. The international community should understand preventing "dangerous anthropogenic interference with the climate system" or "dangerous climate change" as being about poverty prevention. A dangerous

---

34 Shue, 1993.
35 NGO stands for non-governmental organization.

climate policy is one that extends poverty.[36] It is as simple as that.

Hope: Hang on.[37] I agree that poverty eradication should be *a* basic objective. Indeed, that was agreed in Rio back in 1992.[38] However, should we really *define* 'dangerous climate change' in terms of poverty *alone*? Climate change has lots of impacts. It affects the poor, yes, but also the rich and middle class. If a super hurricane hits Florida, or a megafire burns through California, it's not just the poor who suffer. People from all walks of life die, or have their lives upended. I'd say all of that is dangerous. I'd also include the natural world. Climate change that wipes out huge numbers of species of animals is dangerous. "Dangerous climate change" should be understood as whatever covers all of that stuff, not just interference that extends poverty. Suppose we choose a climate policy that did slightly better at eliminating poverty but 200 million more middle-class people died because of it. In that case, I'd say that we weren't successfully meeting our objectives.

Peter: Fair enough. But suppose we get past the name "dangerous climate change" and focus on the substance. We believe that, even though other impacts are relevant, eliminating poverty should take absolute priority over them. Minimizing poverty should be the overriding consideration; only when that is achieved should we focus on other issues. In particular, any energy policy that prolongs global poverty is unreasonable.[39]

Hope: I'm still skeptical. I'm not comfortable giving poverty *absolute* priority over everything else that matters. We

36 Moellendorf, 2014, ch. 1.
37 The objections to Moellendorf below are drawn from Gardiner, 2017a, pp. 441–449.
38 See Principle 5 of the Rio Declaration.
39 Moellendorf, 2014, p. 22. See annotated bibliography.

don't normally think that way. I agree that poverty is very bad for people, but death, disease, and other forms of suffering are bad too. In addition, we usually give some weight to the interests of people who are better off and to nonhumans, even if we give stronger weight to the worst-off. So, a less absolute anti-poverty principle seems needed.

Peter: I disagree. At some point, you just have to dig in your heels and make a stand. It will be a better world if we truly put absolute value on human life, and that means prioritizing poverty.

Hope: Are you really saying that we shouldn't delay poverty alleviation for a single day, even if doing so would save the lives of millions of middle-class people? That's what your absolute priority principle seems to imply. It seems extreme.

Peter: So be it.[40]

A pause, as they each take this in.

Peter: Let's see what we can agree on. I'm guessing that we both think that poverty is a great scourge and believe that eliminating it should be a high political priority demanded by basic justice. Is that right?

Hope: Yes. I agree that failing to address global poverty constitutes a great injustice. But that injustice *predates* climate change. I also suspect that it could be eliminated quickly, and relatively cheaply.[41] For example, some argue that

---

40 Faced with a hypothetical counter-example of the sort that "a billion people not in poverty would be very likely to enjoy improved social conditions, but this would require the low but not insignificant probability that poverty eradication would be postponed by one day" (Moellendorf, 2014, pp. 26–27), Moellendorf bites the bullet, saying that we are "simply committed to the 'unreasonable' outcome" imagined in the hypothetical, and "there is no way to state the commitment to eradicate poverty that will allow for soft limits" (Moellendorf, 2014, p. 27; emphasis added).

41 Pogge, 2002; Singer, 2011, pp. 214–215.

it would take only about 1% of global Gross National Product to eradicate severe poverty.[42]

Peter: So, we agree! But why does it sound like you are opposing my call for climate policy that does not extend poverty?

Hope: Well, it sounds like you are committed to trading off other kinds of climate harms against poverty reductions.

Peter: If necessary. We believe that both climate change and climate policy are threats to poor people. Negative climate impacts can hurt them, but so can climate policies that impede development. That's why we say that any energy policy that prolongs global poverty is unreasonable.[43]

Hope: Hmm. Let's find a way to make this dispute more concrete, so we can see what's at stake. What do you think about emissions targets?

Peter: We believe that climate policy can be done in ways that pass the poverty test. We are in favor of the 2°C limit, since we believe that can be achieved without prolonging poverty.[44] The benefits of staying below 2°C are very high for the global poor, and the economic costs are manageable and worth it. More than 2°C is dangerous.

Richard: That sounds reasonable.

Hope: I agree that going beyond 2°C is definitely very dangerous. But what about 1.5°C? That seems to be a much better target, according to the IPCC. And many developing countries have been pushing hard for it.

Peter: We think the carbon budget for 1.5°C is too tight. It is too ambitious. The economic costs seem very high.

Amina: (*gasps*) What?! I thought I agreed with some of what you've said, but this is outrageous! Why do you think 2°C is acceptable if you care about the most impoverished and

42 Glennie, *et al.*, 2020.
43 Moellendorf, 2014, p. 22.
44 Moellendorf, 2014, p. 24.

vulnerable? If we don't keep warming under 1.5°C, the Maldives will be wiped off the map by sea-level rise!

Peter: That's terrible. All the same, the global poor need access to cheap energy. So, we must protect those emissions. Unfortunately, in lots of places, that still means burning coal. Indira said as much.

Amina: Coal? (*angrily*) That's the dirtiest fossil fuel. We can't possibly have a climate policy that continues to allow coal! It will destroy the climate! It will engulf my nation!

Peter: We might have to accept some environmental damage if we are to protect the poor.

Amina: *Some* damage? (*standing up in outrage*)

Indira: I understand Amina's anger. But we won't get anywhere if we let that run the room. I want to be clear that we don't believe that taking a stand against economic injustice necessitates allowing up to 2°C warming. It may even entail the opposite. My original comment about poverty was merely intended to show that equal-per-capita alone, without any further obligations on behalf of developed countries, is inadequate.

Hope: Amina's and Indira's reactions makes me wonder. What do the global poor actually think about your proposals, Peter?

Peter: What do you mean?

Hope: Well, many impoverished countries have been pushing hard for 1.5°C.

Peter: I'm not sure why. The 2°C temperature limit has been accepted internationally.[45] Don't you see that we're trying to protect the poor countries from disaster?

Hope: It's worth asking whether *we* have the best grip on *their* situation.

Peter: What do you mean? The poverty is pretty self-evident! So is the climate vulnerability.

---

45 Moellendorf, 2014, p. 24.

Hope: Shouldn't you be at least asking them? It seems disrespectful just to *assume* all that.

Amina: I know poverty well. The Maldives has recently done better economically, but my family grew up in destitution. And I can tell you that poverty is not a monolith, nor are the views of the people living under it.

Joseph: I represent the Democratic Republic of the Congo. I see plenty of scientific evidence that 1.5°C is a better target for the global poor than 2°C.[46] It's also not a coincidence that I have yet to speak, since the voices of those representing the worst off are rarely requested.[47] We need to push to be heard, for our situation, values and culture to be recognized.[48]

Amina: Agreed! In resisting 1.5°C and endorsing 2°C instead, Peter, we're concerned that you are giving too much credibility to the voices of the global north and too little to those of the global south. It is not "science" alone that originally determined the 2-degree target. Everyone inside the debate admits that that 2-degree target initially emerged from a political process dominated by the global north and its scientists. The activism of the global south has largely caused the new focus on 1.5°C. In Copenhagen in 2009, my country's president said that with "anything above 1.5 degrees, the Maldives and many other small islands and low-lying islands would vanish."[49] African representatives have insisted that accepting 2°C would

---

46 Lenferna, 2018, ch. 9.

47 The Democratic Republic of the Congo is one of the world's poorest countries, ranked 175th out of 189 on the UN Development Programme's Human Development Index. Some recent effects of climate change on the country are described by journalist Vava Tampa, who goes on to describe the consequences of a 3 degree temperature rise to the Congo as "unthinkable" (Tampa 2021).

48 "Recognition justice requires that policies and programs [fairly consider and represent] the *cultures, values, and situations* of all affected parties" (Whyte 2011).

49 BBC, 2009.

amount to a "suicide pact" for much of their continent.[50] Hence, the slogan "1.5 to stay alive."

Joseph: We are concerned that our voices are not taken seriously enough. Philosophers point out that there can be epistemic injustices, which are injustices surrounding how people's knowledge and views are treated.[51] When it comes to climate, much of the talk is about protecting the poorer and more vulnerable nations, but our testimony is often ignored, and our experiences and concerns are not well understood. Too much is shaped by the words and experiences of the richer nations, who often have little understanding of our situation, our lives, and our cultures.

Hope: Thank you for speaking up, Joseph. Please continue to do so – every country's voice matters in this discussion.

Peter: Fair enough. What do you advise, Joseph?

Joseph: For a start, we should be wary of assuming that scientists and politicians in the developed countries share enough of the life experiences and background knowledge of people in less developed countries to know what's best for them. Besides, some actors may have a vested interest in misinterpreting the concerns of the worst off.[52]

Peter: I agree. We need to move more carefully. I suggest that we talk more after this meeting if you're willing. I'd also like to invite any others interested in poverty to join us.

Many around the room nod, and nobody presses the topic further. Hope feels somewhat relieved.

Hope: Are there other things we've missed so far in this discussion?

Joseph: Well, my country cannot claim to speak for the whole global south or the world's poor. But maybe we can offer our current perspective.

Carlos: Please do!

50   Welz, 2009.
51   Fricker, 2007
52   Fricker, 2007, p. 172.

Part 2: Justice beyond Mitigation

Joseph: We have concerns about how this whole discussion has proceeded. We agree with the consensus that the rich should bear greater burdens, but other issues are being obscured. For one thing, we are troubled by the almost exclusive focus on emissions cuts. We understand the urgency there. But that is not all we need to address. Other things matter too.

Amina: Yes! My country is already feeling climate impacts, and the situation is urgent for us. If our land entirely submerges under the oceans, what will be done? Of course, we don't have the resources to deal with the problem alone. We can't survive hurricanes or sea-level rise without substantial help. The international community needs to assist us. This goes beyond the resources we need to develop more cleanly, through green energy and new technology – though we need help there too.

Hope: Hmm. I suppose we usually assume that much of what we say about emissions reductions will carry over to other areas of climate policy. You know – the general ethical consensus that the richer countries should take on heavier burdens in the transition; the relevance of polluter pays, appropriator pays, beneficiary pays ... that kind of stuff. (*Hope trails off as she senses frustration in the room*)

Carlos: You mean the questions that the rich countries are obsessed with! All those quibbles about whether or not you're responsible, and detailed arguments about the precise nature of your responsibilities. Really, if you're honest with yourselves, isn't it just *blindingly obvious* that you are responsible? It seems to us that you are just trying to find ways to wriggle out of it – or else to keep arguing long enough that you won't actually have to do anything.

Susan: (*trying to rescue the situation*) Maybe it would be help-
           ful to step back and remember that there are different
           areas of climate policy. While mitigation is important,
           other areas are too. Adaptation is one; loss and damage
           is another. Amina and Joseph are right to point out that
           we haven't discussed those yet.

Adaptation and Loss & Damage

Richard: (*trying to help Susan*) I've always been confused about
           something there. What exactly are the differences
           between adaptation on the one hand, and "loss and
           damage" on the other?

Susan: (*looking at her briefing notes*) Well, the IPCC defines
           adaptation as "the process of adjustment to actual or
           expected climate or its effects." They add: "in human
           systems, adaptation seeks to moderate or avoid harm or
           exploit beneficial opportunities" and "in some natural
           systems, human intervention may facilitate adjustment
           to expected climate and its effects."[53]

Marcia: That makes adaptation sound like a very dry, techni-
           cal notion. But it is far from it. Adaptation will be the
           very life-blood of our future, of our survival! How can
           our societies adjust to rising seas, hotter summers, and
           more extreme weather? How can we ensure humanity
           still thrives in a warming world?

Hope: Right. Adaptation is really about our values.[54] How we
           protect and promote what matters most.

Richard: (*teasing*) I know you were a philosophy major; but do
           you really see everything in terms of values …? (*he stops
           abruptly, realizing this comment may be inappropriate
           in the middle of a negotiation*)

Hope: (*unfazed, and with a light laugh*) Actually, yes!

53 IPCC, 2014, p. 114.
54 Hartzell-Nichols, 2014, p. 151.

Indira: I think what Hope means is that the climate impacts that typically matter to us are those which will damage, or otherwise negatively affect, the things we value. So, understanding what we value is going to be central to thinking about adaptation.

Amina: The same is true for loss and damage. I think 'loss' is best understood as the permanent destruction or sudden unavailability of something of value. 'Damage' is when what is valued remains available, but its function is somehow impaired.[55] So, again, values are central. No matter how much we adapt, climate impacts will cause loss and damage.

Joseph: One thing emphasizing the role of values shows is how easy it is for discussions of impacts to get too narrow. For instance, some tend to think of adaptation in purely monetary terms. Others believe the aim is development but see that as simply getting communities affected by climate change back on a Western growth path as soon as possible – where that usually assumes a capitalist model heavily oriented towards consumerism and high energy use.

Richard: I don't understand. Isn't that what everyone wants? To build their economies, accumulate wealth, and escape poverty?

Marcia: It is more complicated than that.

Richard: Enlighten me.

Marcia: Some countries focus on economic development, but not all. We believe one should think of development in broader terms. For example, it is easy to slip into a narrow understanding of prosperity, based on the value of goods traded on markets, and measured by GDP per

---

55 For a philosophical analysis of what is meant by loss & damage, see McShane, 2017. On how it is distinguished from adaptation, see Wallimann-Helmer, *et al.*, 2019.

capita.[56] But this fails to capture much of what people actually value. Also, raw economic numbers tend to make the discussion very abstract very quickly, which tends to obscure what is really happening on the ground.

Richard: Can you flesh that out a little more?

Marcia: Of course. In the policy discussion, it is common to distinguish between economic and noneconomic values. Economic values include much of what you might expect: income and property across all sorts of sectors, such as agriculture, manufacturing, tourism, and so on.

Richard: But you are unhappy with measures such as GDP. What's the alternative?

The Capability Approach

Marcia: There are several possibilities. The most popular framework is the capability approach.[57] It encourages us to think of development in terms of what people can do and be. Prosperity involves genuine opportunities to be healthy, well-nourished, educated, or to do things like live a life of a normal length, move around freely, participate in government, practice religion, etc.[58] This approach is used extensively outside of climate circles: for example, by the World Bank in its Human Development Index.

Richard: How does this help us think about adaptation, or loss and damage?

Indira: One thing is that different communities can support capabilities in different ways, as influenced by history, culture, and community values. So, for example, while mobility – the basic ability to get around in one's

---

56 Gross domestic product per capita. GDP per capita is the measure of the value of the goods and services produced within a country over a certain time period (usually a year), divided by that country's population.

57 For a philosophical survey of the capability approach, see Robeyns & Byskov, 2020.

58 Sen, 1980; Nussbaum, 2003.

community or society – is typically important for everyone, this may look different in different places.

Richard: I get that. New York already has its subway; maybe Los Angeles should get one too. But that approach probably won't work everywhere. My district is mainly rural and spread out; it also gets serious snow. So, I have to think about how my constituents will adapt if we move beyond fossil fuels. Maybe Los Angeles can do away with the pick-up trucks clogging its freeways, but in my state we need good mountain vehicles in the winter. I'm not arguing against change, but I do want change guided by our situation and values. For my state, I'm thinking that turning our heavy-duty mountain trucks electric is a priority.

Amina: I hadn't thought of that example.

Indira: I take it we all agree it is important to get stakeholders involved in climate adaptation, to see what they value and why. If major changes must be made, those affected must deliberate about what priorities to set and how to move them forward. One advantage of the capability approach is that it is more sensitive to these things than more old-fashioned frameworks like GDP.

Noneconomic Values

Amina: This is just as important when we talk in earnest about noneconomic values.

Richard: Such as?

Marcia: Well, with climate, we see threats to culture, health, community, place, relationships to the natural environment ....

Richard: Yes. Hope has been telling me about the effects in our own country, to indigenous peoples in Alaska. They are losing their homes, their abilities to hunt and fish, and many other things that not only impose physical damage, but also undermine their ability to maintain their way of life and survive as a culture.[59]

Marcia: In Brazil, indigenous peoples have much experience with this. We have had our whole way of life deeply disrupted. The white man has moved us around, confined us when we once roamed free, tore us from our homelands, and forced us to live in new landscapes. Then, we are told to "adapt." But often these "transitions" have devastating effects on our communities and on the individual lives of our people. They also ignore our relationships with the forest, the ways in which we see our lives intertwined with those of other creatures and the land itself. A narrow, Western economic picture of adaptation threatens to erase much of who we are.[60]

Joseph: Marcia's point is a good one. It also gives me a chance to introduce my second concern about the way the climate discussion usually goes. I've been holding back, waiting for the right opportunity.

Hope: Please – go ahead.

The Threat of Maladaptation: The Great Wall of Lagos

Joseph: I believe we need to understand climate change, including adaptation policy, against a wider background, including the context of global capitalism, the legacy of colonialism, institutional racism, and so on. If we do not understand the context – the way the world is now and how it came to be how it is – we will not be able to see climate change in the right light, and we will inevitably make mistakes. For example, we may adapt to climate in ways that only further injustice.[61]

---

59 Whyte, 2016. See annotated bibliography.
60 Heyward, 2014.
61 For a primer on the myriad ways adaptation in particular can go wrong, see Schipper, 2020.

Hope: Can you give us an example?

Joseph: Of course. Have you heard of the Great Wall of Lagos?

Hope: No; sorry!

Joseph: It is a very revealing story. The Great Wall is an adaptation project in the Nigerian capital of Lagos.[62] Due to climate change, Lagos is increasingly threatened with sea-level rise, and especially storm surges. In response, the authorities are building a barrier – a sea wall – stretching for 8.5 kilometers along the coast. A new city is being built on top of the wall, called Eko Atlantic. It will have 250,000 residents and is being promoted as a sustainable city that will "pave the way for Nigeria to become a top 20 global economic power within the next decade." Those in charge promise "tree-lined boulevards, manicured gardens, elegant plazas, three marinas and a stunning ocean front promenade" that will be "a complete contrast to the congested, traffic-filled streets of Lagos."[63] It will have marinas, tennis courts, all sorts ....

Richard: Sounds cool!

Joseph: In the abstract, perhaps. But who is it for? Two- and three-bedroom apartments at Eko Atlantic cost between $700,000 and $3 million American dollars each. You are probably not aware, but over 60% of Nigerians – 100 million people – live in extreme poverty; even middle-class households have incomes of only around $15,000–$30,000 a year.

Amina: So, you're saying that this "adaptation" project – this new city – is for the rich, most of whom will likely be foreigners. It is not designed to protect the local population, and especially not the poorest and most vulnerable.

---

62 The proceeding discussion of Lagos draws heavily from the discussion in Tuana, 2019, pp. 6–11. See annotated bibliography.

63 As quoted in Tuana, 2019, p. 7.

Joseph: I should also tell you that the new sea walls are designed to *divert* a storm surge, they *don't prevent it*. So, the water will likely be pushed to the sides, up and down the coast.

Amina: (*shocked*) I fear that you are now going to tell me that this is where the poorer, most vulnerable communities are.

Joseph: I am so sorry; yes, you are right. I must also tell you that many poorer people have been pushed into those areas *by the building of the wall*. The Great Wall, you see, creates valuable new waterfront property. So, on several occasions, the government has forcibly evicted local fishing communities on very short notice to acquire the land, burning their homes as they go.[64] The displaced then moved to areas that are now at heightened risk.

Amina: That is horrible.

Joseph: But not entirely unexpected. Modern Nigeria is a very unequal society, built on petrodollars and the legacy of slavery. It is not surprising that "adaptation" would take the form of enhancing the position of the elites at the expense of the historically oppressed. Indeed, racism and colonialism play a complex role in explaining how this situation has come about. Those being displaced are largely indigenous populations, whereas the elites were formed under the influence of the slave trade and British colonial rule.

Background Injustice and Climate Apartheid

Carlos: Cases like this illustrate why Archbishop Desmond Tutu once warned that adaptation is "becoming a euphemism for social injustice on a global scale," and why he feared "climate change apartheid."[65]

---

64 Tuana, 2019, p. 10.
65 Tutu, 2007, p. 166.

Joseph: Yes. What I want to emphasize is that we keep talking about the poor and vulnerable as if these are independent, fixed categories; yet the whole global system is built on a legacy of great injustice, including colonialism, feudalism, and structural racism. Moreover, the problems are ongoing. The current global economic system is very unjust. It is based on economic and political rules that discriminate against some populations and compromise their efforts to flourish. We cannot understand climate injustice in isolation from all that.

Indira: Agreed. There's a long and ugly history of environmental injustice around the world, where marginalized populations are disproportionately the victims of environmental burdens. Indeed, much of the pioneering work emerged from your country, Hope and Richard. Racism plays a big role – that's just a fact. Consider Triana, Alabama, where toxic pesticides were dumped into local creeks and poisoned the town's black population.[66] More recently, think of the terrible water quality in Flint, Michigan, a majority-black town; or Hurricane Katrina, where minority communities were devastated by the deluge.[67]

Joseph: Our negotiations must recognize this backdrop injustice and refuse to contribute to its continuation. Environmental racism is global in scale, but so is resistance. At the grassroots, communities of color are banding together across borders to face down the threat; we should see our work as part of that.[68]

Hope senses that Richard is getting restless.

66 Bullard, 2000.
67 Young, 2006.
68 Mersha, 2018. For a book-length analysis of reparations for slavery and colonialism — and how they are intimately tied to the climate crisis— see Táíwò, 2022.

Susan: I see that matters are complicated, and that justice is a big issue. What can be done? What are you asking of us?

Amina: Past agreements have promised funding for adaptation, which is good. But the support – the actual money – has been slow to follow. Developed countries have also balked at addressing loss and damage, mostly for fear of admitting liability. But who will pay to rebuild neighborhoods devastated by hurricanes and homes destroyed by fires? Who will pay for the costly relocations of entire communities? (*looks pointedly at Richard and Hope*)

Richard: (*wearily*) I wondered when we would get back to money again.

Amina: (*bristling*) Money is only the beginning; this is about so much more! In my country, the Maldives, as sea level rises, our whole way of life is threatened. If we do not see robust climate action soon, we will literally lose our homeland. We may become a people without a place ....[69] (*Amina pauses for a moment, trying to keep control of her evident grief*) So, financial compensation, though useful in some ways, does not come close to covering what is at stake for us.

Hope: I'm so sorry, Amina. What else would help?

Anima: Well, if the international community were truly serious, it would be thinking about how we can have a place to be. There are various ideas. Perhaps special passports, entitling us to live in a country of our choosing? Or better still, the promise of territory that we could call our own? We would want a place that is in some sense familiar. We are an island people, wedded to the ocean. We do not want to be uprooted inland, to a desert or forest. I know that this will seem a lot to ask to some of you.

---

69 Simona Capisani argues that states have a right to a livable locality, which would be violated in the event of territorial loss due to sea-level rise. This grounds a moral obligation to protect against climate change-induced displacement. See Capisani, 2020.

But remember, none of this is of our doing, and even this would not truly make up for what is being lost .... (*Amina trails off*)

Can't We Just Act on Climate Alone?

Richard: (*with an irritated glance at Hope*) I'm starting to feel overwhelmed by all this. With all due respect to Amina's country and their terrible situation, these perspectives on the problem – concerns about global inequality, colonialism, racism, and so on – threaten to derail our efforts to move forward. Are you seriously saying that we need to solve all the world's problems, past, present, and future in our climate policy? That seems a tall order; it may also be counterproductive.[70]

Amina is visibly shocked; Joseph grimaces

Joseph: (*stiffly*) Unfortunately, you remind me of Martin Luther King Jr.'s *Letter from a Birmingham Jail*. He chastised the white liberal as a person who "paternalistically believes he can set the timetable for another man's freedom" and who constantly advises black folks to wait for a "more convenient season," usually out of fear of white backlash.[71] Telling victims of oppression that their liberation must await the defeat of climate change echoes that.

There is an abrupt pause as people take this in

Carlos: (*intervening*) It would be nice to make the world whole – still, I for one am not holding out for that. However, I don't want climate policy to make things worse – to exacerbate injustice – and I'd like to make things at least a little better. If we can take a step in the right direction,

70   Posner & Weisbach, 2010, p. 98.
71   King Jr., 1963.

it would help establish a baseline of trust that would help here and with other global problems.

Hope: Call me an idealist, but I also see it as an opportunity for reconciliation, a way of starting to build the foundations of a more cooperative and equitable world order.[72] I don't mean that we will achieve some perfect deal. That may not even be possible. There have been too many wrongs, and many cannot be compensated for. Nevertheless, it can't be a mere token effort. It must be meaningful. It should include some level of aid for those who have been marginalized, with the goal of reintegrating them into global society on an equal footing. Even without perfection, the symbolic value of such an effort would be worth a lot.

Richard: I don't know what you mean. Do you have an example of something like that?

Hope: Not a perfect one. But I was thinking of the settlement in New Zealand between the Crown and the Ngai Tahu tribe. Substantial resources were involved – $300 million, I think – as well as some powers of first refusal to buy back land when it became available. No one thought that this was really enough to compensate for past injustices. However, it was a genuine contribution, and the symbolic value was significant as well.

Joseph: I fear you are too optimistic.[73]

Richard: So do I. Hope, I see where you are coming from. But I'm not sure how easy that approach to climate change would be in many countries, including the United States. I also worry that it will open the floodgates. C'mon, Hope, do you really think we could sell that? Do you think the United States is ready?

---

72 Whyte, 2016, p. 14.
73 For a forceful critique of the settlement experience from a Maori scholar, see Mutu, 2018.

Amina: (*before Hope can reply*) Ready? We're supposed to wait for you to be ready?

Richard: (*taking a deep breath*) I didn't mean to cause offense. I just think it might be better, and more feasible, to move on climate alone first. The other stuff is best left for other venues.[74]

Amina: Which venues?

Richard: Well, I don't know. Foreign aid or trade policy ...

Joseph: Hmm. Could you honestly commit the United States suddenly to become much more sympathetic on aid and trade if we deal with climate independently?

Richard looks down at the table and remains silent.

Joseph: (*sighs*) That's what I thought.

Marcia: Your suggestion also misses the point: it's not possible to "act on climate alone." We *can't* ignore background injustice. Climate presents an all-too-familiar scenario: black, brown, and indigenous peoples will be forced to move and adapt not because of something they did, but because of the actions of white people. It's *déjà vu*: similar to many other instances of colonially induced environmental change.[75] That's why we *must* actively consider these histories of oppression and the silent workings of power when we choose our mitigation targets, what methods we use to pursue them, how we adapt and who we adapt for, and how losses and damages are handled.[76]

Joseph: Well said, Marcia. It seems the results of recent climate negotiations cannot be explained *without* racism on the part of wealthy countries, whether conscious or

---

74 Posner & Weisbach, 2010, p. 88. For an extended argument to this conclusion, and a consideration of objections, see ch. 4 (especially pp. 79–98). For criticism, see Caney, 2014; Gardiner 2016b; and Gardiner & Lawson, 2021.

75 Whyte, 2017, p. 155.

76 Tuana, 2019, p. 4.

unconscious, to some extent. African countries know this well. The pledges taken after Copenhagen would not be enough to buy us coffins.[77] Taking background injustices seriously is a first step to ensuring recent history doesn't repeat itself.

Indira: For all these reasons, Richard, I'm afraid we must insist that nothing gets off the ground without *meaningful* progress on clean energy finance, development, adaptation, and loss and damage. Until you show us a commitment to these things, after all, we cannot be sure you really have our best interests at heart, and that your climate policy is not serving as cover for further exploitation.

Richard: (*winces*) I can understand your point of view. But I am not sure my party or my country can deliver all of that.

Joseph: I know. That's a big part of why we're so concerned.

Richard: (*abruptly*) Then it seems we are at an impasse ....

Hope: Wait! That's not true. I think we're making progress.

Richard: (*continuing, in a clipped, professional tone, but with a sadness in his voice*) I am happy I attended this meeting. I learned a lot about other's perspectives. But, unfortunately, I see a lack of pragmatism at play here. The perfect is the enemy of the good, after all. If the representatives in this room keep digging in their heels – demanding the burden fall disproportionately on one country or another and refusing to compromise – we will remain stuck in place until climate tragedy washes us all from our various moral high grounds.

Amina: Moral high ground? (*angrily*) We feel the water lapping at our feet! We are appealing to your own moral principles to help prevent my country's total inundation. If the entire world doesn't come together and take aggressive action, there will be no land in my country high enough

---

77 To paraphrase a leading climate negotiator, as quoted in Hari, 2009.

to retreat to. We no longer have the privilege to compromise on half-measures.

Carlos: (*intensely*) If anyone is digging in their heels here it's you, Richard. Even your fellow representative is more reasonable.

Richard: (*stands up*) I'm sorry, but I see no point in continuing. No amount of our quibbling will achieve the perfection so many of you demand.

Hope: Nobody is under the illusion that perfection is possible, Richard. Perfection hasn't been possible for a long time.

Richard: (*stiffly*) Thank you all for your time. (*leaves abruptly*)

Indira: I'm afraid he's right, Hope. I know you thought that we could come to solutions our country's leaders could not, but it is not happening. We are retreading the same discussions, with identical outcomes.

There is a general murmur of agreement, and others begin to stand up.

Hope: Please, everyone. You understand how important this is.

Carlos: Thanks for your effort, Hope (*walking over*). I think we made progress here, even if it may not seem so.

Joseph: Yes; thank you, Hope. For what it is worth, my country has rarely had the platform to voice our perspective to other countries like I just did. (*smiles weakly, holding out a hand. Hope takes it*)

Hope: I am glad to hear that. (*returning the smile*)

Before long Hope sits in an empty room. She doesn't get up. Instead, she remains there for another half an hour, staring at the empty chairs and reflecting on the meeting. Eventually, she rises with a sigh, pulls on her coat, and steps out into the cold.

Maybe there is still time to arrange another meeting before they all go their separate ways. Maybe there is still time to fix this mess.

Dialogue 5   A Big Technological Fix?

Shortly after the previous meeting

Part 1: Geoengineering, Some Basics

Hope nervously looks on as Richard checks his watch beside a projector screen, tapping his foot. Hope is surprised not to have been consulted about this new meeting he has arranged for the junior negotiators. She thought they were in this together. Nevertheless, she welcomes the opportunity to clear the air, engage again with the others, and maybe lay some groundwork for moving forward.

Richard: Thank you all for coming. As you know, I felt our previous discussion got off track. I'd like to try again, this time focusing on a different approach. It is one I find more productive, optimistic, and forward-looking; one that emphasizes our common interests, rather than our differences.

Richard pauses as if waiting for affirmation, but the room remains silent.

Richard: What many of us are trying to achieve is economic prosperity, for our countries, for ourselves, and for future generations. Obviously, technology will play a big role. We must transition to cleaner forms of energy, develop new power plants, systems of transportation, and so on. But I believe we also need to think bigger ...

Kim: We agree ...

DOI: 10.4324/9781003123408-5

The others are taken aback by this sudden intervention from the usually-recalcitrant Chinese representative. Any sign of agreement between the superpowers demands close attention. Some wonder whether the exchange has been pre-orchestrated.

Kim: (*continuing*) Despite our increasing emissions, my country still has much poverty. But we also have a thriving technological industry. We'd like to look to this sphere for efficient solutions to this great climate problem. Some approaches may not require us to endure the extreme economic harms of drastically cutting our emissions over only a few decades.

Indira: (*surprised*) What do you have in mind?

What Is Geoengineering?

Kim: Are you familiar with geoengineering proposals?[1]

Indira: Yes. "Geoengineering" is a pretty general and contested term. But I suppose you are referring to grand technological interventions on a planetary scale into fundamental earth systems, with the aim of counteracting climate change? I believe the United States spends a small portion of its research budget investigating it.[2]

Kim: These possibilities interest us too.[3]

Murmurs around the room.

Richard: (*seizing the initiative*) Let's begin our discussion by bringing in some experts. This is Ernest, a physicist-turned-engineer, and Thomas is an expert on Earth systems.

---

1 For overviews of the ethics of geoengineering, see Gardiner 2019c, Callies 2019, Pamplany, et al., 2020.

2 In 2020, U.S. Congress approved $4 million dollars to subsidize geoengineering research. Fialka, 2020.

3 In 2020, China revealed plans to develop a "weather modification system" over a slice of land the size of India. Griffiths, 2020.

Ernest: Thank you, Richard. "Geoengineering" is most commonly defined as "the deliberate large-scale manipulation of the planetary environment to counteract climate change."[4]

Thomas: Some propose a broader definition that uses the term for large-scale, intentional manipulations of the planetary environment *for any purpose*.[5] That might become important in the future, as human impacts on the Earth continue to accelerate. But, as Ernest says, in the current moment "geoengineering" is mostly used for interventions that counter climate change. "Climate engineering" is another way of referring to them.

Ernest: Numerous techniques are being discussed, some very different from each other. There's some controversy about precisely which count as "geoengineering." Some even doubt that the word picks out a distinct category at all.[6]

Thomas: In our view, the controversy is not surprising, given that the definition contains terms which require further clarification, like "large-scale" and "deliberate." Nevertheless, we suggest that we don't allow that problem to prevent us getting started.[7] After all, definitional questions are common in other areas, including those involving science, policy, and law. In law, for example, "reasonable doubt" presupposes some idea of "reasonable"; and in policy "cost–benefit analysis" requires specifying what counts as a "cost" or "benefit."

Ernest: Two broad categories of geoengineering techniques dominate current discussion.[8] The names are fairly self-explanatory. The first – carbon dioxide removal (or CDR) – involves taking carbon dioxide out of the

---

4 Shepherd, *et al.*, 2009.
5 Keith, 2013; Blomfield, 2015, p. 42; Gardiner, 2016a.
6 Heyward, 2013, pp. 23–24; Jamieson, 2013, pp. 529–530.
7 Gardiner, 2016a.
8 Keith 2013, XVIII.

air. That directly reduces the greenhouse effect by tackling its cause. There are many possible strategies that might be attempted. Everything from planting more trees, to enhancing natural systems like weathering, to developing machines to suck carbon dioxide directly out of the air.

Thomas: The second kind of technique is solar radiation management (or SRM). SRM strategies aim to reduce the heating caused by increases in greenhouse gases by deflecting away some quantity of incoming sunlight. If less energy comes in from the Sun, the Earth will not heat up as much. Again, there are numerous possible techniques, from painting roofs white, to making clouds brighter, to pumping sulphate particles into the stratosphere, to putting mirrors in orbit. All would attempt to increase the planet's albedo – its reflectivity – and so send more radiation back into space.

Richard: Thank you. I propose we start with CDR. Any objections?

Differing Visions

Hope: Before we dive into specific technologies, I think it would be useful to discuss *why* we might pursue them at all. That may shape our whole approach.

Richard: Isn't the "why" obvious? We need every tool we can get our hands on to combat climate change. The global situation is already worrying. It may soon become alarming. It makes sense to investigate the full portfolio of potential solutions – every tool in the toolbox.

Indira: Hope, I too am puzzled by your question. If geoengineering may provide us with opportunities to "buy time" to transition away from fossil fuels, surely it is worth pursuing. We believe that CDR is very important, in part because it may give us longer to develop our economy.

Peter: May I jump in? We see SRM as potentially a way to protect the global poor against the worst climate impacts.

Maybe the only way, at least in the short term. A compelling Plan B. How could anyone object to that?

Kim: We also believe SRM might be relatively cheap and easy compared to decarbonization. For such reasons, stratospheric sulfate injection looks particularly interesting.

Richard: Right. Importantly, it can work politically. Unlike mitigation, SRM does not entail structural overhaul and significant economic burdens. Therefore, it's more likely to find bipartisan support. My political party will be much more willing to support cheap, free-market-friendly investments in geoengineering. I imagine many countries will feel the same.

Amina: We are interested in these techniques too. Obviously, SRM would be used only as a last resort, but my country may go underwater. So, a last resort may, ultimately, be what is needed.

Carlos: (*with a slight chuckle*) Hold on. I see what Hope is getting at. Already these appear to be very different rationales for geoengineering.

Richard: Really? And anyway, why does that matter? If everyone has some reason to support it, isn't that a good thing?

Carlos: I'm not sure. The reasons may pull in different directions. For example, arguing that geoengineering might be "cheap and easy *compared to decarbonization*" suggests that these technologies are being pursued as an *alternative* to emissions reductions, at least in the near term. However, that doesn't gel very well with claims that the situation is so desperate that we should be deploying "every tool in the toolbox" – that suggests that geoengineering is being *added* to aggressive mitigation, as part of enhanced ambition.

Marcia: I'm also puzzled. The "last resort" framing implies we're planning for an emergency deployment, under extreme conditions when we've *already tried everything else*. But "buying time" suggests getting started as soon as possible, so as to ease the transition *while we do the other things.*

Carlos: I'm not saying the different arguments are necessarily incompatible. In principle, perhaps the various tensions might be reconciled. Still, I have a suspicion that in practice political agreement will not be easy. So, I wonder about the wisdom of proceeding without further discussion of our aims.

Joseph: And who are the "we" in this discussion? With all due respect to Peter, I find the idea that SRM will be deployed by or on behalf of the global poor a little far-fetched. Isn't it more likely that rich countries, or the elites, will use it to protect their own interests, and to wriggle out of their mitigation responsibilities?

Carlos: I agree. Moreover, even if protecting the poor were the agreed aim at the start, I'm not sure who I'd trust to implement such a policy and maintain it over time.

Hope: More generally, I'm worried that, taken in the abstract, the idea of geoengineering risks being everything to everybody. This sometimes happens with new technologies. People see a world of possibility. They then fill that world with their own hopes and dreams for the future. But usually life is not that simple. Some dreams are unrealistic; some come into conflict; and sometimes one person's dream is another's nightmare.

Marcia: I'm also worried about a politically convenient optimism bias. We often see inflated promises made on behalf of new technologies. Early advocates for nuclear power promised electricity "too cheap to meter." They meant power so inexpensive that companies would not find it worthwhile to charge people for how much they used. Obviously, that didn't happen.[9] Also, nuclear power brought with it new problems, such as risks of accidents

---

9 Smil, 2016.

and what to do with radioactive waste that remains dangerous for thousands of years.

Every Tool in the Toolbox?

Susan: Hmm. This discussion is interesting. But, Hope, what about Richard's original argument that we need to investigate the full portfolio of potential solutions – every tool in the toolbox? Why is that controversial? We all agree that the situation is getting grave, especially for some communities.

Joseph winces at the word "grave."

Hope: The portfolio argument is too quick. Notice that many possible tools to reduce climate risk are not seriously under consideration. For instance, I don't see us talking about radical proposals for cutting consumption, or controlling global population growth, or rejecting meat. Yet all of these would make a big difference to the climate. Why is geoengineering *automatically "in"* the portfolio when other unconventional proposals seem *automatically out*?

Susan: Those other ideas are very radical.

Joseph: And geoengineering proposals aren't? Attempting to take control of the climate system seems pretty radical to me! Injecting sulfates into the stratosphere, for example, is downright scary ...

Susan: Point taken. Still, some of the alternatives you mentioned raise moral objections. Population control, for instance. That has a troubled history. Surely, we should steer away from those kinds of controversies.

Hope: I agree that ethical concerns need to be discussed; indeed, I think this should happen across climate policy. But notice that some have serious ethical reservations about geoengineering too. My point is that it begs an important question *simply to assume*, without argument, that geoengineering techniques are automatically in the policy portfolio *just because they might help* with the climate problem, whereas other things that almost certainly would help are assumed to be out, again without argument.[10] A more even-handed approach wouldn't prejudge the issue for or against one kind of approach. It would consider the ethical pros and cons of all the options.

Richard: So, you would not wish to prejudge the issue *against* geoengineering, right?

Hope: Of course not. I merely don't want to prejudge in its favor either.

Indira: One caveat. This idea that we should be pursuing *all* the things that *might* help? It seems too permissive a standard. We don't have the time or resources for wild goose chases. Some of these geoengineering ideas sound *very* speculative. We should focus our energies on solutions that stand a *reasonable chance* of working.

Richard: Agreed. ... This seems a good time to get more specific. (*others nod*) Let's review some of the more promising technologies, and then turn to general issues about geoengineering. Naturally, today's discussion will be preliminary, but with luck it will set the stage for a longer-term engagement. Ernest ...?

Ernest: Certainly. I'll begin with CDR – techniques that removes carbon dioxide from the air. "CDR" is quite a large umbrella term. It includes a range of technologies that differ significantly in their implementation and risks, from planting trees to direct air capture, to enhanced weathering, to biotech.

10   Gardiner 2010, 2019c.

Thomas: Notice that not all instances of removing carbon from the atmosphere *count* as geoengineering, properly speaking.[11] For example, planting a few trees in your neighborhood is not sufficient. You'd need to plant a huge number to make a meaningful difference on the planetary scale.

Afforestation

Richard: I assume no one has a problem with planting trees?
Carlos: Actually, Marcia and I worked together on a fact-finding exercise a few years ago for Latin American countries. Based on that experience, I believe we need to move slowly.
Marcia: I agree. While most people would probably support planting a few trees, or even a modest number of forests, the scales being talked about here are a different matter. For instance, some scientists have floated the idea of storing 752 billion tonnes of CO_2, or *two-thirds* of human-made carbon emissions, by adding 0.9 billion hectares of forest.[12] That is a huge amount of land – almost the size of two Amazon rainforests. Moreover, much of the land that would need to be converted is highly valuable to many communities. Advocates of afforestation – tree planting on "non-treed land"[13] – sometimes seem to forget this. Large-scale afforestation inevitably leads to land conflict, and local populations tend to lose out.[14] I prefer protecting the forests we have. Respecting the land rights of indigenous peoples

---

11   Blomfield, 2015, p. 43.
12   Bastin, *et al.*, 2019. These estimates provoked significant pushback in the scientific community, and were accused of making multiple mistakes. Marshall, 2020.
13   https://archive.ipcc.ch/ipccreports/sres/land_use/index.php?idp=48
14   Pichler, *et al.*, 2021; Marshall, 2020.

is an excellent way of doing this in my country of Brazil.[15]

Ernest: Protecting existing forests is essential, of course. Nevertheless, to make a meaningful difference to climate change, we really are talking about a massive intervention. Genuine geoengineering: CDR that alters the climate on a planetary scale.

Carlos: This worries us. Who gets to decide? We do not like the idea of the rest of the world claiming the forests as theirs, to do with as they please. We see the forests as *our* forests. They are not the property of the rest of the world. We should decide. After all, many of you developed countries cut down your forests centuries ago. Perhaps that was your right. But it does not give you claims on us now.

Richard: Clearly there are issues to be addressed, but we cannot expect any climate solutions to be completely cost-free.

Carlos and Marcia exchange unhappy glances.

Bioenergy (BECCS)

Indira: (*hurriedly*) Perhaps we should turn to other methods of carbon dioxide removal. Which seem most promising?

Ernest: So far, the most commonly discussed is "bioenergy, combined with carbon capture and storage" (BECCS), usually pronounced "Becks." It is the CDR technology the IPCC emphasizes most in its models.[16] What makes BECCS so attractive is that, in principle, it can remove CO_2 from the atmosphere while also *producing energy*.

Thomas: BECCS involves three key stages.[17] First, biomass is grown, derived from products like sugar cane waste or fast-growing tree species like willows, which draw CO_2

15 Baragwanath & Bayi, 2020.
16 Baragwanath & Bayi, 2020.
17 Consoli, 2019, p. 4.

from the atmosphere as they grow. Second, this biomass is burned for usable energy at a conversion facility, producing CO_2. Third, this CO_2 is captured and stored. The aim is that the overall outcomes will be a net reduction in CO_2 in the atmosphere. So long as the CO_2 taken out during biomass production is greater than than any lost to the atmosphere during the burning for energy, BECCS produces usable energy while drawing down CO_2.

Richard: A system that produces energy while removing greenhouse gases from the atmosphere! Doesn't that sound perfect?

Indira: *Amazing!* I'm intrigued.

Ernest: BECCS' potential is enormous! Some call it a "savior technology."[18] For instance, it might revolutionize hard-to-decarbonize industries, such as cement production.[19]

Thomas: Hmm. I'd counsel caution. So far, the development of BECCS has been disappointingly slow.[20] Moreover, there are challenges with compiling and transporting vast quantities of bioenergy.[21]

Carlos: The biggest obstacle may be one BECCS shares with afforestation. Implementation at the scales the IPCC assumes would require *massive* amounts of land, perhaps between 0.4 and 1.2 billion hectares, or 25–80% of current global cropland.[22]

Marcia: Good point! Notice that, unlike afforestation, BECCS gobbles up *arable* land. Instead of growing food that could be used to feed the hungry, that land will now be used to grow biomass to be burned. But this suggests increasing food insecurity, as *both* BECCS and climate change will be putting pressure on food supplies,. The

---

18   Hickman, 2016.
19   Hickman, 2016, p. 7.
20   Consoli, 2019, p. 4.
21   Anderson & Peters, 2016b.
22   Fajardy, *et al.*, 2019.

cruelty in that should be obvious. Famines result from the poorest people being priced out of the market as food supplies dwindle.[23] Again, we face the worry that the so-called "savior" solution may sacrifice some to protect others. That seems a great injustice.

Ernest: I take your point. Avoiding injustice when implementing BECCS requires walking a tight-rope. We must take enough land to stabilize the climate, but not so much as to cause malnutrition.

Joseph: Excuse my skepticism, but I have no confidence that our current global system can deliver that degree of surgical precision.[24] Indeed, it seems foolhardy to rely on it.

Indira: I now question whether BECCS is such a savior after all.

Direct Air Capture (DAC)

Richard: Maybe this is a good time to turn to direct air capture.

Earnest: DAC is a chemical process where CO_2 is captured directly from ambient air and stored in geological storage sites. DAC takes far less land than BECCS or afforestation, and so shouldn't result in the same kind of moral trade-offs. It has one-thousandth the land intensity of BECCS.[25]

Thomas: Note that DAC is distinct from carbon capture and storage (CCS), which is usually about capturing carbon from industrial processes on site, when it is in concentrated gaseous form. Due to the carbon in ambient air being less concentrated, DAC takes significantly more energy than CCS. This is one reason why direct air capture is a lot more expensive than other forms of CDR, and than mitigation.[26]

---

23   Sen, 1983.
24   Shue, 2017, p. 207.
25   Gambhir & Tavoni, 2019, pp. 406–407.
26   Gambhir & Tavoni, 2019, p. 407.

Marcia: This is usually the biggest objection to DAC. There are also concerns that some forms will require significant amounts of clean water.[27]

Ernest: That being said, there is considerable scope for innovation and cost reduction due to the technology's newness.[28]

Richard: Precisely! The technology is still young, but DAC has huge potential. Ultimately, I consider it to be one of the most promising ways that CDR could be deployed at scale, and in an ethical way.[29] After all, it would basically just be a form of pollution control.[30]

Amina: Even if you are right, do we have time to wait for DAC?

Ernest: It is hard to say, but more research can surely help.

Stratospheric Sulfate Injection (SSI)

Richard: Shall we turn to the other kind of geoengineering, solar radiation management?

Ernest: Yes. As we said earlier, "SRM" technologies seek to reflect solar radiation away from the planet, to cool the Earth down. Something as simple as painting rooftops white can count, and some scientists are interested in whitening clouds over the oceans.[31] But the most exciting idea is stratospheric sulfate injection (SSI).

Richard: (*to the room in general*) Did you know that volcanoes can cool the Earth? The eruption of Mount Pinatubo in the Philippines in 1991 spewed smoke into the higher reaches of the atmosphere that dampened down global temperature for a couple of years.

27 Preston, 2013, p. 31.
28 Preston, 2013, p. 31.
29 Peacock, 2021, p. 14.
30 Preston, 2013, p. 24.
31 This latter idea is known as marine cloud brightening (MCB). See Latham, *et al.*, 2012; Wood & Ackerman. 2013.

Ernest: The idea to use SSI to combat climate change takes inspiration from this. We would inject reflective particles into the stratosphere in order to reflect some sunlight back to space, and so cool the planet down. If we keep the surface below the 2°C or 1.5°C thresholds, this would hold off many of the negative effects of global warming, such as extreme weather and rising sea level.[32]

Richard: This technology is probably the *only* thing that could cool the planet quickly, in a few years, if temperature rise started to get really out of hand. It is also quite likely to enhance the productivity of ecosystems, and even increase crop yields.[33]

Indira: Intriguing!

Richard: That's not even the best part: the underlying science is sound and could be deployed in just a few years for a few hundred million dollars – the cost of a Hollywood blockbuster! [34]

Kim: There is much promise, I can assure you.

Joseph: Wow. This is really radical thinking. Still, something about it doesn't sit right with me. For example, the claim that it would be cheap seems misleading, even myopic. You seem to be focusing only on the immediate costs of *deployment* – of spraying stuff into the stratosphere. I'm not sure those are the "costs" that matter most. Isn't it a little like me claiming that it would be cheap to do brain surgery on Peter just because I can buy a good scalpel for $20? Shouldn't we be much more worried about *what happens* when I cut into his head? Especially if I don't know much about brain surgery, and plan to learn on the job![35]

Peter: (*wincing*) Geez, Joseph. You'll give me nightmares.

---

32 Keith, 2013, p. x.
33 Keith, 2013, pp. ix–x.
34 Keith, 2013, p. ix.
35 Gardiner 2010, 2019b.

Joseph: Sorry. But it is nightmares I'm worried about here.

Indira: Ernest, can you say more about side effects? Wouldn't SSI be the same as deliberately polluting, like putting a layer of smog permanently above us?

Ernest: There may be some negative effects.[36] The sky may no longer be blue; plant health and photosynthesis could decline. The effects of large injections on the stratosphere itself, and on the biosphere as a whole, are unknown.

Marcia: Wait. The ozone hole is in the stratosphere, and it's just starting to close up again. How would it be affected?

Thomas: We don't know yet.[37]

Indira: That's troubling.

Marcia: I also remember learning that the eruption of Mount Pinatubo decreased precipitation and brought drought to some parts of the world.[38] Would SSI cause that too?

Ernest: (*irritated*) The science is still developing. Obviously, further investigation would be needed.

Kim: We cannot know all the effects for certain. However, we think that this technology should be researched so that if we need it, we know as much about its impact as possible. [39]

Hope: Perhaps. Still, you've got me wondering ... Doesn't that speak against the idea that we could deploy within just a few years? I take it there is a big difference between merely being capable of spraying stuff into the stratosphere, and actually deploying SSI in even a *minimally responsible, ethical* way. Surely there would need to be

---

36   Robock, 2008.

37   There is some reason to believe that sulfuric injection would damage the ozone layer. This has led some to suggest injecting a different aerosol, such as calcite (Keith, *et al.*, 2016). Scientific debates around the best aerosol to inject has led some to start using the more general term stratospheric aerosol injection, or SAI.

38   Jamieson, 2014, p. 220.

39   Cicerone, 2006, is one prominent advocate of this idea.

a huge amount of research and testing before we should even consider deployment?[40]

Thomas: Certainly. We must approach the science responsibly.

Hope: How long do you think that would take?

Thomas: At least a few decades. Personally, I'm skeptical we'd be able to recommend deployment with any confidence before mid-century at best.[41] It may take considerably longer.

Indira: Really? This makes me skeptical about counting on this technology to buy time for us to decarbonize.

Thomas: There are deeper issues too. Some scientists argue that ultimately there is no way to do adequate "field testing" without effectively *deploying* full-scale geoengineering.[42] Others emphasize that we have only one planet to experiment on, and that is already changing rapidly due to climate change. As a result, we cannot perform anything like a clinical trial – there is no control group.

Ernest: Maybe it will be easier to test than it initially seems …

Peter: May I say something? We support SSI because we're deeply concerned that mainstream mitigation and

---

40 Lenferna, *et al.*, 2017.

41 Thomas' attitude is based on our assessment of the current literature. For instance, on the optimistic end, in 2013 David Keith estimated that sensible deployment would not be possible before 2035 (i.e., 22 years from 2013), assuming testing started in the early 2020s (Keith, 2013, p. 88). However, since research has been slower than Keith hoped, and testing has yet to begin, we suspect that even that optimism would now (2021) push deployment out to the mid-2040s. In a more pessimistic assessment, Alex Lenferna, *et al.* claim that the obstacles for ethically relevant testing are significant. For instance, they "question whether climate response tests of regional impacts can be undertaken in a way that is scientifically and ethically relevant for a deployment scenario," since such tests are likely to take too long, or require too large a radiative forcing, to help with any ethically reasonable evaluation process (Lenferna, *et al.*, 2017, p. 582). Given this, presupposing a research timetable that results in ethically responsible deployment by mid-century seems optimistic. (Full disclosure: Gardiner is a co-author on this paper, which emerged from a course he co-taught with a leading climate scientist, Tom Ackerman.)

42 Robock, *et al.*, 2010.

adaptation plans are not ambitious enough. Most countries aren't making the needed sacrifices now, even the easier ones. Clearly, countries should mitigate first, but geoengineering seems like a promising Plan B.

Carlos: We need to be careful.[43] If I understand Ernest correctly, most scientists are not thinking of SSI as *an alternative* to emissions reductions. They are assuming that we *have to do both*: that mitigation is still absolutely necessary even if SSI comes online. So, in a sense this alleged Plan B *presupposes the success of Plan A*. I'd like to emphasize that this is not how we normally use the phrase "Plan B." It's not, I think, how Peter is using it either. Maybe what we should say instead is that SSI is essentially in a new Plan A+. It is a way of adding to mitigation and adaptation, not replacing them.

Marcia: There's another reason against labeling SSI as Plan B. That implies that it is the second thing we're trying, right after Plan A. But we've been working on the climate issue for decades now. I think we're already way beyond Plan A or Plan B. Plan F maybe.

Peter: Fair enough. Maybe I should chose my words more carefully. I guess my view is that SSI is a promising *back-up plan*, in case mitigation and other things fail.

Hope: A back-up plan *would* be handy. But I'm not comfortable assuming SSI can play that role. In fact, I wonder whether it is too early to be calling pursuing that technology a *plan* at all. It seems way too speculative – too "pie in the sky" – and I have no idea whether we could govern it successfully internationally. As Joseph said earlier, the whole idea may end up being just wishful thinking.[44]

43   This discussion is based on Fragnière and Gardiner, 2016.
44   Some believe the prospects to be so bad that SSI should not be pursued at all (e.g., Hamilton, 2013; Hulme, 2014).

Amina: Right. SSI sounds too dodgy to serve as Plan B, or a genuine back-up plan. A "last resort" then, as I said earlier? Probably high risk, and low probability of success. A "Hail Mary pass", as you Americans like to say.

Richard: Perhaps. Still, better that than nothing, right? Sometimes you win with a Hail Mary!

A thoughtful silence ensues.

Richard: (*continuing*) Thank you, Ernest and Thomas. The initial overview has been helpful. I propose we open the floor for wider discussion.

Justification & Context

Hope: I wonder if it would be helpful to distinguish three questions. Sometimes I'm asked: "are you for or against geoengineering?" This strikes me as a bad question.

Ernest: Why?

Hope: The "for or against" framing divides people into opposing camps from the outset, perhaps without good reason.

Richard: Go on.

Hope: On the one hand, probably no one is *unequivocally for* geoengineering, whatever the technology, the circumstances, and state of our knowledge. For instance, I'm betting few would favor deploying SSI right now, given the risks and how little research has been done.[45]

Ernest: Agreed.

Hope: On the other hand, I doubt that many are *unconditionally against* geoengineering. For example, most probably wouldn't rule out reforestation on a moderate scale, if it could be done in a socially and ecologically sensitive

---

45 Notably, some are proposing trying other kinds of regional climate engineering soon to protect specific areas, such as to counteract loss of Arctic sea ice cover (Desch, *et al.*, 2017; Field, *et al.*, 2018) or the rapid deterioration of the Great Barrier Reef (Tollefson, 2021).

way – if the issues around land, for example, could be resolved.

Many nod.

Susan: Interesting. If the for-or-against question is unhelpful, what would be a better way of setting up the debate?

Hope: Earlier I worried about the "everything to everybody" problem: that people are prone to fleshing out the implications of new technologies in terms of their own hopes and dreams. So, I propose starting with the *justificatory question*: under what conditions would geoengineering become justified? It might prevent us talking past each other.

Thomas: Agreed. Scientists need to know more about what global society is expecting of us. For instance, what threats would trigger an intervention, what we should be trying to protect against, what level of uncertainty society is willing to tolerate, and so on.

Susan: Politically, we also need to understand what governance structures would be needed to manage SSI. What kinds of institutions, laws and policies ...

Carlos: True. But let me emphasize that we see ethical questions as central here. For instance, any intervention should satisfy norms of justice and political legitimacy.

Maria: Yes! We're especially concerned about responsibility. How do we decide who would be in charge? How would they be accountable to the rest of us? How would compensation be handled for negative side effects?

Richard: I would have thought justification would be easy. Surely, geoengineering benefits everyone.

Ernest: Indeed – our modeling shows that SSI would benefit all regions.[46]

---

46 E.g., Smith & Henly, 2021, pp. 4–5; Ricke, *et al.*, 2013. Cf. Gardiner, 2019c, p. 72.

Thomas: Well, actually we have to be careful not to overreach. Most of the modeling so far investigates what happens if one reduces solar radiation uniformly across the Earth: in other words, turns down the sun. It is not surprising that this would reduce temperatures across the globe; but it is some distance from there to saying that *our spraying particles* into the stratosphere would have the same effects. For instance, we don't know how our spraying would interact with the system; in fact, we're not sure how to do the spraying yet.

Hope: Ernest – When you say SSI would benefit everyone, what do you mean? It seems unlikely that any intervention would have only positive implications for *every single person* on the planet, over hundreds of years.

Ernest: Modeling of dialing down the Sun projects fewer changes to temperature and precipitation than severe climate change. That's what I meant by benefiting everyone.

Hope: But I don't suppose that would benefit *absolutely everyone*, right? Those wanting to open up the Northwest passage to shipping, for example, might prefer higher temperatures.

Ernest: Hmm. Okay.

Indira: I'm also not sure that we can use lower changes in temperature and precipitation as a straightforward proxy for benefit. For example, suppose both SSI and severe climate change cause changes in where it rains, but SSI causes less of a change. If it now rains only 300 miles from where it used to, instead of 700, that rain probably still doesn't reach my crops. So, it is unclear how the smaller shift benefits me.

Richard: I think you're all being too picky! We're trying to avoid a catastrophe after all, a catastrophe that threatens poor countries most of all! That seems to make the justification easy. We don't need perfection. So long as geoengineering improves things, it is justified.

Indira: I worry that makes justification *too* easy.

Richard: What on earth do you mean?

Indira: Well, if one is imagining a genuine catastrophe – millions, perhaps billions who suffer or die from climate change – almost anything seems better by comparison. If genuine catastrophe is the baseline, then geoengineering might look justified, but then so might much else.[47] Global dictatorship, for example, or mass sterilization, or all manner of things.

Richard: That's outrageous! No one is proposing such measures! Least of all me.

Susan: (*intervening*) Hope. You said there would be *three* questions ...

Hope: Oh. Thanks for remembering! As well as the justificatory question, we must also ask the *contextual question*: how *relevant* are whatever justificatory conditions we identify *to the world we live in*, or one that may plausibly emerge in the foreseeable future?

Ernest: I'm not following. Can you give an example?

Hope: Sure. Suppose we ended up agreeing SSI could be justified in a world where we have a huge amount of scientific knowledge and appropriate testing, where the global order is more-or-less just, where robust compensation measures are in place in case things go wrong, where all nations and marginalized populations endorse the specific plan ...

Carlos: I see the issue. That world seems pretty distant from our situation. It is also very bold to assume that it will evolve over the next few decades.

Joseph: Hope's point, I take it, is that a good answer to the justification question might not really address the contextual question ... That SSI *could* be justified from some idealistic point of view might be good news for scientific and political geniuses who also happen to be angels. But it is not such good news for us.

Marcia: Angels wouldn't have allowed this crisis to develop in the first place! They'd have started decarbonizing sooner, and be able to pursue better ways out even now. So, any solutions – including geoengineering solutions – should pay some attention to the problems that got us where we are today.

Amina: (*impishly*) We probably shouldn't be planning something that could *only* be successfully implemented by *Madame Virtue Incarnate,* the wildly popular, democratically elected, Earth President![48]

Joseph: (*joining in*) *Especially* if it is much more likely to be implemented by His Imperial Majesty, *Corruption Incarnate*! I'm imagining a tyrannical dictator running a superpower, whose main agenda involves global domination and wiping out populations he sees as "inferior."

Carlos: Given our context, even researching some of these technologies worries us. For instance, nobody likes to feel like their time and money has been wasted.[49] So, funding research may dramatically increase the likelihood of SSI actually being deployed, whether or not it is a good idea. I don't like the idea of sleepwalking into geoengineering.[50]

Marcia: Yes! We're not convinced yet that we'd *ever* be in favor of deployment.

Joseph: I'm already uncomfortable with giving too much attention to these "grand, planetary-scale interventions" – this whole session today, for example. I'm especially worried about the whole "savior technology" idea. Many people would love to see some "silver bullet" for climate,

---

48 This example and the next are taken from Gardiner, 2019b.
49 Gardiner, 2019b, p. 6.
50 Jamieson, 1996; Gardiner, 2010, pp. 349–352; McKinnon, 2019.

especially one that allows them to continue on much as before. Again, I fear wishful thinking.

The Threat of Wishful Thinking

Indira: Indeed. We don't have these technologies in hand. Some are pretty fanciful. Isn't it irresponsible to put so much stock in them? We should not pursue a global climate policy predicated on magical thinking. Too much is at stake. We may dream of "carbon unicorns,"[51] and be pleased if they ever show up, but we ought not to make plans that *rely* on them.

Carlos: This may be a good moment to point out that this is happening already. Remember the IPCC's projections that underlie the call for net zero by 2050, and a 45% cut by 2030, if we are to respect the 1.5°C target?[52] Those projections *already assume* a massive amount of BECCS or similar CDR technology.

Joseph: (*shocked*) You're joking, right?

Carlos: Alas, no. Mainstream projections already presuppose that BECCS will work, that it will ramp up in the next decade, and that over the rest of this century it will remove carbon from the atmosphere on a scale that roughly matches contributions from emissions reductions.

Ernest: Yes. Of the scenarios modelled by the IPCC, all 76 that are consistent with a likely chance of not surpassing the 2 and 1.5° targets assume the widespread roll-out of negative-emission technologies.[53] (*Changes slides, and points a blue laser at his slide show*) As this chart shows, we're talking about removing an amount of carbon comparable in size to our remaining carbon budget.[54]

51   Carton, 2020; Friends of the Earth International, 2021.
52   Anderson & Peters, 2016b, p. 183.
53   IPCC, 2016, p. 57; Anderson & Peters, 2016b, p. 182.
54   Anderson & Peters, 2016b, p. 182.

Joseph: (*angrily*) That's outrageous! BECCS is still a speculative technology! We do not yet know whether it works effectively, or can be scaled up to the levels assumed. Why on Earth are we already banking on it? That's shockingly irresponsible. Does everyone see the huge moral problem here? The blatant injustice?

Thomas: Some scientists have complained that such assumptions are morally indefensible. They say the models are making a high-stakes gamble, risking the lives of billions on a bet that the technology will come through in time.[55]

Marcia: This gamble affects every country. If it is lost, many will suffer, especially in vulnerable populations, but everywhere else too.[56] Even now, as climate change intensifies droughts in Africa, it also burns communities in California to the ground. Sea-level rise imperils the Maldives, but also looms over New York, Shanghai, and many other coastal cities.

Richard: Hold on. I'm not sure why you think that pursuing geoengineering counts as a high-stakes gamble, or why you are assuming "gambling" makes things worse. The remaining carbon budget is pitifully small, so we are already in very serious trouble. Surely, putting resources into geoengineering technologies increases the chances of meeting our targets? Consider an analogy. Throwing a life-preserver to a drowning victim may not guarantee a successful rescue. But it is not a "high-stakes gamble."[57] It's an effort to help in an already desperate situation.

Carlos: I don't accept your analogy. Part of the point is that, when we're talking about geoengineering, we don't have a life-preserver yet! And we don't know with any real confidence that we can make one, or get it to the river in time!

---

55 Anderson & Peters, 2016b, p. 182.
56 Anderson & Peters, 2016b, p. 182.
57 This borrows the analogy and wording of Lackner, *et al.*, 2016, p. 714.

Joseph: The analogy bothers me too. Our current behavior with BECCS seems more like knowingly letting someone jump into a raging torrent, telling them we *may* be able to save them, but not making clear that this will depend on a technology we have yet to develop.[58] I see that as a great injustice.

Carlos: I have a more general concern about wishful thinking. If people think we have a get-out-of-jail-free card in our back pocket, will they make the sacrifices needed to mitigate appropriately?[59] Won't speculating about geoengineering serve as a dangerous distraction from the central tasks of mitigation and adaptation? I've heard people say geoengineering poses a moral hazard. It encourages emitting in a similar way that having insurance encourages riskier behavior because the insured feel protected from consequences.

Amina: Hmm. Richard, are you assuming that geoengineering would be deployed *in addition* to conventional climate policies, especially emissions reductions? That the two will not influence one another? I am not so sure. Won't some see even the prospect of geoengineering as a *substitute* for emissions reductions? If we can offset emissions with CDR by pulling CO_2 out even as we put it in through fossil fuel use, or if we can smother the effects of greenhouse gases with SSI, then doesn't that reduce the incentive to cut back, or at least to do so as quickly?

Richard: Perhaps; but it could go the other way.

Susan: Yes! Personally, I find the prospect of SSI particularly scary. Knowing that some are contemplating it makes me *even more* determined to cut emissions.

---

58 This borrows the analogy and wording of Anderson & Peters' reply to Lackner, *et al.*, 2016. See Anderson & Peters, 2016a.

59 Jamieson, 2014, p. 224.

Carlos: The phrase, "moral hazard" is used in numerous ways.[60] As Susan said, often it is associated with a *possible, psychological effect* of having *genuine* insurance. However, what I was emphasizing just now was something a little different: an *actual, policy effect* based merely on the *potential* that CDR might help. People are already accepting mitigation targets based on the *hope* that geoengineering will come through.

Marcia: I feel we must underline a further "moral hazard" – moral corruption. Scientists might *say* that geoengineering is only needed to "flatten the climate curve" while other changes come in, but we see little sign of those other changes right now.[61] We're suspicious that it is all a smokescreen. Perhaps the only thing some countries are really willing to do is geoengineering, whatever you tell us; maybe even whatever you tell yourselves.

Ernest: No serious scientist thinks that! Without decarbonization, SSI would have to continue essentially forever, masking a bigger and bigger temperature rise. But perpetual SSI is unsustainable and irresponsible.

Thomas: Agreed. Still, it is good to raise the issue of substitution. Some scientists do argue that successful SSI would make mitigation less urgent: that it would – even *should* – shift the cost–benefit calculation away from emissions reductions to some extent, and so slow down the pace of decarbonization.[62]

Carlos: As I suspected …

Ernest: I don't agree with that!

Richard: I see there are complications. Nevertheless, I still believe we should research geoengineering, and hopefully

---

60 Hale, 2009. While its meaning varies, the term does usually have moral connotations. However, Benjamin Hale denies that there is anything inherently *moral* about the notion of a moral hazard.

61 For an overview and visual illustration of the concept of "flattening the climate curve," see Aldern, 2020.

62 Keith, 2013, pp. 132–133.

eventually deploy. I say that even if there is a genuine moral hazard – that is, even if doing so reduces the incentives to cut emissions a little. Preparing a life-preserver is morally preferable to not doing so, even if it reduces the person's incentive to learn to swim.[63]

Hope: (*reluctantly*) I guess I'm not sure that adding geoengineering research into the mix makes things *much* worse than they already are. It is not as if we have been doing well reducing emissions so far anyway.[64] There are so many problematic forces at work.

Many look down at their notes, grim expressions on their faces.

Necessity

Richard: (*exasperated*) Look. I've listened patiently so far. And I understand many of the concerns. Still, we must acknowledge the elephant in the room. If our aim is to achieve the temperature targets – those required by climate justice, as you all so eloquently insisted last time we met – then geoengineering is *simply necessary* and at very large scales.[65] That's why it is included in so many IPCC scenarios.

Marcia: You speak as if meeting the temperature targets *physically requires* geoengineering. That's not true. We'd have a very good chance of keeping under even 1.5°C if we stopped all anthropogenic carbon emissions tomorrow.

Ernest: My apologies. I should make clear that the IPCC scenarios are based on what they take to be *reasonable* projections of emissions moving forward, making assumptions about economic growth, population, and so on.

---

63   Here we offer a related analogy to Lackner, *et al.*, for a slightly different purpose (Lackner, *et al.*, 2016, p. 714).
64   Gardiner, 2013a.
65   Lenzi, 2021, p. 7.

Marcia: So, there's already a set of beliefs in the background about what's "reasonable," and what would be "too radical" to contemplate?

Ernest: In a sense.

Carlos: So, the judgments here are ultimately ethical and political.

Marcia: Do the projections include other radical measures, such as consumption restrictions, population controls, or a consensual global lockdown, or anything like that?

Ernest: Some presuppose a very fast transition.

Richard: Please! We must work within the parameters of what the current generation is likely to accept. Ignoring the silly fantasies, the CDR projections are very reasonable. Also, realistically, we will probably *have to* employ SRM at some point as well.

Kim: Recall that SSI in particular may be the *only* thing that is potentially fast-acting enough to hold off a climate catastrophe coming in the next few decades.

Richard: As I said, if we're serious about confronting the climate problem, geoengineering is morally necessary.

Carlos: I'm sorry. I'm suspicious of these appeals to necessity and urgency.

Susan: Do you doubt the gravity of our situation?

Carlos: Not at all.

Richard: What then?

Carlos: Doesn't the appeal to necessity seem a little, how should I put it, morally schizophrenic?

Amina: What do you mean?

Hope: It's a philosophical term. "Schizophrenia" means "divided mind" in Greek. Moral schizophrenia involves a conflict between one's behavior and one's values.[66]

Marcia: Hold on. I'm uncomfortable with using the word "schiz-ophrenia." That's most commonly employed in medical

---

66 Stocker, 1976; Gardiner, 2013a.

contexts, to refer to a specific condition. I worry that using it here stigmatizes people who are clinically schizophrenic. We don't want to encourage the impression that being neuro-typical is the only good way of being human.[67]

Carlos: Absolutely. That was not my intention at all. Perhaps we should call it a form of severe moral dissonance instead.[68] All I want to communicate is that one mark of a good life is that it involves a harmony between one's reasons and how one acts. When these pull apart, it shows a malady of the spirit, a state of alienation from one's values, maybe a kind of self-deception about who one really is.

Amina: I see what you're driving at. How is that relevant here?

Carlos: I'm worried about a particular kind of moral dissonance. Let's call it "creative myopia."[69] Suppose someone invokes a set of strong moral reasons – say, saving humanity from severe harms – to justify a given course of action – say, SSI. But this course of action is supported by these reasons *only because* the agent has already ruled out a number of alternative courses of action – say, emissions reductions, adaptation assistance, lifestyle

---

67 Elizabeth Barnes has influentially defended the "mere difference" view of disability against the "bad difference" view (Barnes, 2016). Barnes explicitly restricts her analysis to physical disabilities (Barnes, 2016, pp. 3–5). She does this, in part, because her defense of the mere difference view rests on first-person testimony of disabled people, something that can be more difficult to obtain reliably from the psychologically and cognitively disabled. However, she leaves open that the mere difference view may extend to the psychologically and cognitively disabled, and other philosophers have argued that it should (e.g. Kaposy, 2018). Moreover, many people with psychological and cognitive disabilities *do* speak reliably for themselves. In the case of schizophrenia, examples include Saks, 2008 and Wang, 2019.

68 For a related use of "moral dissonance," see the business ethics literature (e.g. Lowell, 2012).

69 Gardiner, 2013a.

change – more strongly supported by the same reasons. Moreover, this ruling out is due to motives she has that are less important, and are condemned by the background moral reasons. So, in our case, maybe the motive is the desire to cling to high-end luxury goods, whereas the background moral reason is saving humanity from climate catastrophe. This kind of moral dissonance is really disturbing to me. It seems really duplicitous and reeks of self-deception.

Amina: So, you think the necessity argument is only selectively engaging with the urgency of combating climate change. We insist on it when it comes to geoengineering, but it is not on display elsewhere, for other parts of climate policy.

Carlos: Yes. There's a deep disharmony here. On the one hand, we're emphasizing that the situation is so desperate that we must back a truly radical, high-risk solution that might not work and does not yet even exist; yet, at the very same time, we are not doing so many other things that we know would work, and do exist.[70]

Richard: So? Personally, I'm not worried about "internal harmony." Sounds a bit self-indulgent to me. I just want to get things done. And surely we all agree that something *must* be done.

Hope: Wait. I see Carlos' point. It's as if we're asking ourselves a deeply paradoxical question: "What *must* we do, morally speaking, given that we *have not, are not, and will not do* what we *must* do, morally speaking?"[71]

Amina: Hmm. It's not, perhaps, an incoherent question.

---

70 Gardiner 2013. Gardiner illustrates the moral problem with the example of Wayne's Folly (Gardiner, 2013a, pp. 22–24). Due to space constraints we have not included that here, but instructors may find it useful for class discussion.
71 Gardiner, 2019c, p. 74.

Hope: Perhaps not; but it is *jarring*. Also, it makes me wonder about the moral force of the *must*.[72]

Richard: (*rolling eyes*) I see we're back to the moral philosophy again. Let's take a break.

Part 2: Intentional Climate Change & the Anthropocene

Richard: Talking to others in the break, I came away with the impression that many agree with me that CDR is not very controversial, since in principle it is just a form of pollution control.[73] By contrast, they consider SSI and other large-scale SRM much more concerning.[74] So, I wonder if we can focus on those technologies now.[75] (*Carlos and Marcia look annoyed, but remain silent*) Perhaps people could begin by raising their concerns.

Intention

Hope: One notable thing about SSI is that it amounts to *intentional* climate change. Regular anthropogenic climate change is not, at least not in the same way. We are producing regular climate change as a side effect of our pursuit of other things, like energy and consumption. We are not aiming at it. But geoengineering involves a direct attempt to control the climate ... That makes a moral difference.

72   Gardiner, 2013a, p. 22.
73   Some suggest that the amount of controversy over CDR is counterproductive (e.g., Táíwò & Buck, 2019). For a brief overview of principles for CDR within a *just* climate policy, see Morrow, *et al.*, 2020.
74   This view is common. Against it, Gardiner argues that some forms of CDR are similar to worrying forms of SRM, and that the preference for CDR may itself be moralized and infused with environmental values (Gardiner, 2011a, pp. 344–345; Gardiner, 2011b; Gardiner, 2011c).
75   Principles have also been offered for SRM, including the Oxford Principles (Rayner et al., 2013) and the Tollgate Principles (Gardiner and Fragnière, 2018).

Richard: Why? Surely what's important is what happens on the ground, how many lives are affected. If SSI makes things better for people, isn't it automatically better morally speaking?

Hope: That's true on some views of ethics, but not most. Ordinary, commonsense morality would reject it, for example. Take a simple case. Suppose you're driving home, and you hit a child with your car. The child dies.

Amina: How awful!

Hope: Yes. You might take the view that it makes no difference to the child what my intentions were. The poor child is equally dead either way.

Richard: That's my instinct.

Hope: But we normally think intention makes a huge difference, right? It is one thing if it is an accident, but another thing entirely if you *deliberately* run her down.

Indira: (*shocked*) Absolutely!

Hope: So, it seems more things matter than brute consequences, right? Intentions, for example.

Richard: I see what you are getting at, but SSI isn't anything like that. We're not going to target people for death!

Carlos: We can only hope not.

Joseph: His Majesty, Corruption Incarnate might.

Richard scowls.

Hope: (*hurriedly*) Point taken, Richard. But even setting that aside, we might still have to choose who experiences certain bad impacts. Suppose, for example, that those deploying SSI must choose between one intervention which is bad for the Indian monsoon, and another which increases drought in Africa. In choosing, they'd bear some responsibility for who actually suffers. If they choose the first option, for instance, the extra deaths in India would flow from their decisions. That's not neutral; it seems to matter from the ethical point of view.

Indira: Hmm. Can you say more?

Hope: Sure. Let's try an example that's not about geoengineering.[76] Imagine a wildfire near a large city. The local authorities can't contain it. Their options are to abandon some parts of the city, or open a dam upstream to put the fire out. The first will mean many homes are destroyed by fire; the second, that a small town near the dam is obliterated by the water.

Peter: Tough situation.

Hope: Some think they should invoke the following moral principle: the Doctrine of Doing and Allowing (DDA). It says that "the moral constraint on harming others applies primarily (or most stringently) to *doing* harm to others, and only sometimes (or less stringently) to *allowing* harm to befall others."[77] In this case, the idea would be that in choosing to let the fire burn, the government would be *allowing* harm to occur, but in choosing to open the dam, it would be *doing* the harm. So, according to the DDA, there's a strong presumption for allowing the fire to burn. At least, the damage avoided by opening the dam would have to be very large to justify the moral wrong of destroying the homes near the dam.

Peter: Interesting. I take it that the relevance to us is that SSI is a "doing." It involves initiating intentional climate change. By contrast, regular anthropogenic climate change is seen as more like an "allowing."

Susan: I'm not convinced. The regular anthropogenic climate change – as you put it – seems like doing too. Aren't we equally responsible for both? In the climate case, we're the ones causing both the wildfire and the flood. So, why is this doctrine of doing and allowing even relevant?

---

76   Morrow, 2014. The example is adapted from Foot, 1984, p. 183.
77   Morrow, 2014, p. 129.

Joseph: I'm not sure the causal role is the same. Deploying SSI seems more direct, at least for the authorities involved in doing it. It's more like opening the dam.

Hope: Maybe another principle is better: the Doctrine of Double Effect (DDE). It says: "other things being equal, it is morally worse to bring about a bad state of affairs as an *intended* effect of one's action (or as a means to one's intended effect) than it would be to bring about that same state of affairs as a *foreseen but unintended* side effect of one's action."[78] Here the idea is that the effects of the wildfire are a *foreseen but unintended* side effect of our actions – namely, of burning fossil fuels. But the harms of the flood would be an *intended* effect of our actions – namely, of opening up the dam.

Peter: I see why someone might think that, but I'm not sure. Why isn't the loss of the town due to initiating the flood also a "foreseen but unintended side effect"? Presumably, we'd be glad to divert the flood around the town if we could. So, we don't directly intend the destruction.

Hope: Good point. A key controversy about the DDE concerns what counts as part of the intention, and whether the distinction is strong enough to carry so much moral weight. A lot of people reject the DDE because of this controversy. Still, the principle is taken seriously in many settings. That's my real point here. I'm not trying to defend the DDA or the DDE. I'm merely illustrating why people might think that things other than consequences, in this case intentions, are relevant to geoengineering and other aspects of climate action.

Who Decides?

Indira: This is all very interesting, but just think about the governmental nightmare sulfate injection would be. Which of us decides when to deploy global-scale engineering

78   Morrow, 2014, p. 132.

that will affect different countries and citizens differently? Surely unilateral deployment would be illegitimate. No one gets simply to impose their preferred climate on the rest of the world.

Amina: (*affronted*) You forget my country faces total destruction. I think the Maldives would be justified in deploying SSI if that's what it took for us to live. Who could blame us?[79] We'd be within our rights. It's *self-defense*, which is well established in international law.

Joseph: You seem to assume it would be *the Maldives* or a country like it that would deploy the technology. Do you seriously think that the major powers would permit us to take control of the global climate? Even to protect ourselves? Does your attitude change if it is China, or Russia, or the United States? That might be more realistic.

Richard: The United States would not accept any deployment of SSI outside of our control.

Kim: Excuse me. My country could not accept American stratospheric injection. It would have to be Chinese.

Indira: Interesting. Each of you is assuming that you'll control the technology. I wonder if your enthusiasm wanes if you relax that assumption. You can't all be right.

Peter: (*winks*) If we're thinking of deployment later in the century, who knows who the superpowers will be by then? Maybe Brazil, India, or Iran will do it? Maybe North Korea?

Indira: I'm not sure about *that*. But I also worry more generally about the implications of appealing to self-defense. I imagine other countries also believe they have a right to self-defense against you. Might they think that right is triggered by your climate intervention? We could end up with a geoengineering arms race! That would be

---

79 This paraphrases Nathan Myhervold, as quoted in Specter, 2012, and discussed in Gardiner 2013b, on which some of the following is based.

awful for everyone, perhaps worse even than the climate change that gets it going.[80]

Amina: I hadn't thought of it that way. We're not likely to win any arms race, including a geoengineering one.

Susan: (*hurriedly*) Perhaps an international body should make the call.

Joseph: That seems more promising. But comprised of whom? Controlled by whom? And how would it get the authority it needs? Would the powerful nations reject it as a threat to their sovereignty? Also, could this international body be held *liable* for the droughts or extreme weather caused by the climate they created?

Richard opens his mouth, and then closes it again.

Hope: Look, between its speculative nature, uncertain effects, moral controversy, and borderline ungovernability, I think sulphate injection is a far-fetched solution right now. We need to get our heads out of the clouds and focus on mitigation, adaptation, and perhaps less radical forms of climate engineering, like carbon removal.

The Lesser Evil?

Richard: (*sighs*) I think we need a sense of proportion. SSI might be risky, scientifically or politically. But you're ignoring that it might be needed, to protect all of us, and especially the global poor. I think we have a moral obligation to the global poor to research and probably eventually to deploy. Compared to climate catastrophe, SSI is still the lesser of two evils.

Amina: I sympathize. We should not assume that SSI must be perfect, especially as a last resort. Too much is at stake. A heroin addict may need hospitalization, therapy, and rehab; but if they refuse, methadone is better than death

80 Gardiner, 2013b.

by heroin.[81] The globe is addicted to fossil fuels, and if we refuse to give them up, then geoengineering is better than climate catastrophe. A catastrophe my country faces more than most.

Kim: I agree. We must *arm the future* with the technology, so that if they find themselves facing down an impending climate catastrophe, they have the option to risk SSI instead.[82] In your flood and fire example, I think it would be wrong to deprive the city of the *choice* to save themselves at the cost of a district. Likewise, we owe it to the future to ensure they have the option to engineer the climate, even if we are giving them an uncomfortable and morally complicated choice. If we do not, we will be responsible for the suffering and deaths that they can no longer avoid.

Joseph: (*grumpily*) Those are clever metaphors, but metaphors aren't arguments.[83]

Marcia: I wouldn't be too quick to assume that SSI would be the *lesser* evil. If it goes wrong, physically or perhaps politically, SSI might turn out to be *worse* in the long run even than enduring a climate catastrophe. For instance, it might shut down the monsoon, destroy the ozone layer, or provoke another world war. I'm not arguing that these things will definitely happen. But it seems morally irresponsible simply to *assume* that they won't.

Richard: Such will be the factors future governments must weigh.

Hope: I'm also not sure about the phrase "lesser evil."[84] Some think of evils as things that ought not to be done. That would rule out SSI if it were an evil, even a lesser one.

Richard: I guess I was assuming a weaker sense of "evil" – something we'd normally try hard to avoid, and would have good reason to resist.

---

81 Here, Amina paraphrases the environmental scientist Stephen Schneider as quoted in Nerlich & Jaspal, 2012.
82 Gardiner 2010 spells out and analyzes the Arm the Future argument.
83 Jamieson, 2014, p. 219.
84 The following discussion draws on Gardiner 2010, 2012, and 2017a.

Hope: SSI is at least that. But I wonder if there's another sense of evil that is relevant. Perhaps SSI is not merely something we'd normally try to avoid, and so do reluctantly; but it's also not something that ought never to be done under any circumstances, however desperate.

Carlos: What do you have in mind?

Hope: I'm thinking of cases where there's a terrible choice to be made. We may think that it ought to be made; but even if we do, the fact of having to make it *tarnishes* or *mars* the agent's life. I mean it makes it less choice-worthy from the ethical point of view.

Richard: I'm not following.

Hope: A classic example might be *Sophie's Choice*. In the novel, Sophie is taken to a concentration camp with her two young children. She's offered the chance by a Nazi doctor to save *only* one of her children, but only on the condition that she chooses which is to be saved.

Indira: That's a horrible situation.

Hope: It is. I also think it was a very serious moral wrong to put Sophie in that situation.

Indira: What happens?

Hope: Sophie does choose. But she feels terrible about it, and for years afterwards. I've memorized her words, they struck me so: "in some way, I know I should feel no badness over something I have done like that. I see that it was – oh, you know – beyond my control, but it is still so terrible to wake up these many mornings with a memory of that, having to live with it. When you add it to all the other things I've done, it makes everything unbearable. Just unbearable."[85] (*Hope pauses*) In other words, she sees how the choice might be justified, that it is not her fault that she was in the situation, that she probably did the right thing in one sense. But it is still horrible for her, and something she struggles to live

---

85 Styron, 1979, p. 538. As quoted in Gardiner, 2010.

with, being a mother who chose between her children. Her choice weighs on her so much that she eventually commits suicide.

Indira: That's very sad.

Richard: But what does it have to do with our argument about geoengineering?

Hope: Well, someone might say that choosing her son was an evil, but a lesser evil. It was in some sense justified, even perhaps what she ought to have done, morally speaking. Still, the fact that it was an evil means something for Sophie. She did nothing wrong; she should not be blamed. Nevertheless, having done it casts a shadow over her life. It mars it from an ethical point of view. No one would want to be Sophie.

Indira: Interesting. So, I guess your thought is that this might carry over to geoengineering too, right? That even if choosing SSI is the lesser evil, even if you think choosing it would be morally justified in some nightmare scenario, even if it is in some sense what one ought to do and one shouldn't be blamed, nevertheless it might *mar* the lives of those who do it.

Hope: I also think its morally serious to impose such a choice upon future people or governments. I worry we'd be like the Nazi doctor. He inflicts a special kind of wrong on Sophie. We might be guilty of inflicting a special kind of wrong on the future, even as we are allegedly "arming them." And this would be in addition to other wrongs.

Amina: I find Sophie's predicament very moving.

Recognition and Domination

Marcia: This seems a good time for me to let you know that I've been finding much of our discussion problematic. Alienating, in fact. I feel like it does not recognize my people and our circumstances. That seems like an injustice.[86]

Hope: Oh no! What are we failing to recognize?

Marcia: There's much talk that SSI might be done to protect the global poor and indigenous peoples like mine from devastation. But from an indigenous perspective, devastation has already occurred. In fact, it is ongoing.

Richard: What do you mean?

Marcia: In many places around the world, peoples have had their whole cultures and ways of life upended by colonialism, imperialism, and the whole economic and geopolitical system. Indigenous peoples especially have been impacted, often with severe humanitarian costs. In some ways, climate change is just the next threat, building on and intersecting with all of that.[87]

Richard: That is genuinely tragic. Still, I'm not seeing your point. Surely you're not opposing something that might actually help you now, just because of past mistakes?

Joseph: *Mistakes?* I think you mean deep injustices. Sometimes *horrific* injustices ...

Amina: I also think "tragic" is the wrong word. It makes it sound like a simple misfortune, something that was beyond anyone's control. But that's not true. There are deep historical forces here, building out of colonialism and racism.

Richard: Sorry. I don't mean to minimize that history.

Carlos: Thank you. Still, you'll forgive us if we're a little skeptical about your new-found interest in protecting us, especially through a powerful technology which you control and which is being developed without our input.[88]

---

86 As mentioned in Chapter 4, "recognition justice requires that policies and programs [fairly consider and represent] the *cultures, values, and situations* of all affected parties." See Whyte 2011, Martinez 2014..

87 Whyte, 2018, p. 297.

88 Hourdequin, 2019. See annotated bibliography.

As a friend of mine put it: "there has been too much horrible stuff done to us in the name of science to trust geoengineering."[89] Yet, you unflinchingly pose such a solution to perhaps the greatest problem of our time.

Ernest: I assure you we scientists are trying to protect you!

Carlos: (*skeptically*) All of you?

Marcia: I've heard these kinds of arguments before. We live and breathe these things. Your promises lack credibility. As Nemonte Nenquimo, from the Ecuadorian Amazon, says: "When you say that the oil companies have marvelous new technologies that can sip the oil from beneath our land like hummingbirds sip nectar from a flower, we know that you are lying because we live downriver from the spills."[90]

Joseph: I'm not sure that good will from scientists would matter much anyway. Ultimately, they won't be the ones calling the shots. Of course, it is always *possible* to paint a pretty picture of what your geoengineering will look like – a fairytale of justice and protecting the vulnerable. But we've heard such stories before. The reality tends to be a nightmare – for us especially. Perhaps we'd rather deal with a changing climate than political superpowers trying to seize control of the whole system.

Richard: So, you would rather face drought and death than trust another country with the solution?

Joseph: I don't think you appreciate that SSI comes with a hefty price tag. For one thing, it is risky. But even if it works, you'll be taking control of the whole climate system. That puts in place *another* tremendous concentration of power over us, not just now but probably for hundreds

---

89 This quote is taken from interviews done by Christopher Preston and Wylie Carr of native populations from the Solomon islands, Kenya, and Alaskan. See Preston & Carr, 2019.
90 Quote from Nemonte Nenquimo, cofounder of the indigenous-led nonprofit organization Ceibo Alliance and one of *Time* magazine's 100 most influential people of 2020. See Nenquimo, 2020.

of years in the future. Once again, we'll be subjugated. Think about that. Are you surprised we're concerned?

Richard: I hadn't thought about it that way.

Carlos: With all due respect, perhaps it is because you are used to being privileged that you simply assume that these big technological fixes will be implemented in ways that protect and benefit you. We don't have the luxury of making that assumption. Our whole experience speaks against it.

Richard: Even if it is the only way to avert catastrophe?

Joseph: If it is the only way, it is because you've made it so. The most vulnerable communities account for a tiny amount of carbon emissions, and many of us would already be in compliance with very strict limits.

Richard: I hear you. I just don't see an alternative to SRM if things start to get nasty. And you know how vulnerable your country will be to runaway warming.

Carlos: You might be right that we'll end up having no choice but to submit because the alternatives are all so bad. But that shouldn't make you feel good about yourselves. You caused this whole mess, you didn't clean it up, and now you're saying that we have to submit to you even more. Perhaps more drought and death is coming anyway, some of it at your hands. If we do agree it will be with heavy hearts and great anxiety. You are putting us in a difficult position. Our consent will be tainted. We do not like to be complicit in our own oppression.[91]

Joseph: We're tired of being under your thumb. Acting like you are our saviors while you perpetuate the problem.

A tense silence falls.

Worldviews

Marcia: May I intervene?

Richard: Please do.

---

91 Gardiner, 2013b.

Marcia: The indigenous communities in Brazil have lived in the Amazon for thousands and thousands of years, and we've seen our home changing. The rains now do not come until November or December; the river runs low. The deep roots of our trees are torn asunder for pasture. All the while, our indigenous communities – and especially our indigenous women – are kept out of the decisions that so greatly affect our home. The citizens of my country do not expect you to save us. We are asking for your respect. We wish you to listen to what we have to say.

Richard: Fair enough. What do you want to say right now?

Marcia: (*smiling*) Thanks for asking. The biggest thing is that my people feel you have lost your way. You do not understand what it is to live with the forest, nor the deep love and reverence that it teaches. You are in trouble, and your trouble is a threat to every form of life on Earth. Yet still you try to force your civilization and your solutions on us.[92]

Hope: Please, go on.

Marcia: There's a basic difference in outlook. You think of the Earth as something there for you to manipulate for your own ends, as if it were a tool or machine to bend to your will. Sulfate injection is just another example of your instrumental, dominating attitude to nature.

Richard: What's the alternative?

Marcia: Our cultures and traditions are bound up with a different vision of the relationship between humans and nature. We see ourselves as one among many in the natural world. The birds, the animals, the rivers, the mountains. These are all our kin. When you say, "let's intervene in the system and try to take control of it," it violates those relationships.

Richard: Okay. But if it's *necessary* to prevent catastrophe …

92 Nenquimo, 2020.

Marcia: You keep saying it is "necessary" and seem to think yourselves justified, without much of a second thought. We see it as wounding Mother Earth, and our brothers and sisters.[93]

Richard: That is a different perspective.

Marcia: I doubt you really understand. It violates our sense of self. It is partly through these relationships and their history that we know who we are. They are part of our very identity. We see ourselves as belonging to the natural world, rather than having ownership over it.[94]

Richard: I'm trying to listen, I really am. Look, I wish I could see a better way out. But geoengineering seems like the only feasible way forward. What could our countries do to change your mind about it?

Marcia: I'm not sure you can.

Carlos: I appreciate your question, Richard. Some feel like Marcia. But there are multiple perspectives in our countries, and others are more open to talking about geoengineering. I suggest that, to engage properly, you might start by demonstrating your respect. Make us part of the conversation in a real way. Show you accept our right to self-determination, and our different worldviews. Then, we may be able to move forward.

Richard: How might we do that in concrete ways?

Indira: I suggest beginning by showing your commitment on the other things – mitigation, adaptation, loss, and damage. Without that we cannot be sure you really have our best interests at heart.

---

93 Such a perspective is offered by Hands Off Mother Earth! (H.O.M.E), a grassroots campaign against geoengineering. See their *Manifesto Against Geoengineering* (Hands Off Mother Earth!, 2018).

94 For example, this idea is signaled in the philosophies of the Māori people of Aotearoa/New Zealand (e.g., Watene 2016, pp. 292–293) the Lakota on the North American continent (see Standing Bear, 1933, p. 205) and the Yanomami in the Northern Amazon (see Inoue, 2018, p. 26), to name a few. See also Cuomo, 2021.

Amina: Perhaps, if you're so keen on SSI, and think it is abso-
           lutely necessary, you should help us to build *our* capac-
           ity to do it; even put us in charge. We want something
           we can actually take home and get engaged with.[95]

Anthropocene

Marcia: Something else bothers me about SSI.
Amina: What's that?
Marcia: We've been calling it *radical*. That's understandable and
           common.[96] Still, etymologically, radical means *root*. A
           radical solution, then, is one that gets to the bottom of
           the problem. SSI doesn't. On the contrary, it seems an
           extreme logical extension of the domineering attitudes
           that got us into this mess in the first place.
Joseph: Even the name 'solar radiation management' assumes
           some kind of control is needed, that we have the neces-
           sary expertise, and ultimately that we have the right to
           do it.
Richard: I don't think there is any alternative but to try to right
           the ship, so to speak. Humanity has *already* so impacted
           the natural world that geologists are considering declar-
           ing a new, human-dominated geological epoch – the
           Anthropocene.[97] So, some kind of management – or
           whatever you want to call it – seems inevitable at this
           point. If there is to be an answer to the problem of con-
           trol, it's going to be more control.[98]

---

95 This language is adopted again from an interview conducted by Preston
& Carr, 2019, p. 317. Patrick Taylor Smith also argues that political legitimacy
requires putting SSI research into the hands of people in the developing world. See
Smith, 2018. See also Rahman, *et al.*, 2018.
96 Flavelle, 2020.
97 Crutzen, 2002, p. 23.
98 Lynas, 2011; Ellis, 2015; Kolbert, 2021, pp. 6–8.

Kim: I agree. Climate change has now affected every aspect of the biosphere.[99] Other influences abound: microplastics have moved into virtually every crevice on Earth.[100] The temperature, the rain, wind, the very content of the earth we stand on – all of it is being altered by humans.[101] Why get fussy now?

Marcia: Doesn't this bother you? Aren't you worried about turning the world into something artificial? We seem to be arriving at "the end of nature."[102]

Carlos: Sobering, isn't it? The prospect of a fully humanized world runs counter to thousands of years of religious philosophy and traditional belief systems worldwide.[103] Now, geoengineers wish to take this a step further, to *deliberately* seize control of nature's basic operations, allegedly as a means of avoiding the consequences of existing environmental exploitation.[104]

Marcia: Its foolhardy. Perverse, even.

Richard: Once, I might have agreed. My fondest childhood memories occurred in America's National Parks. These experiences instilled in me a love for wilderness. However, I've grown to realize that this love affair was just a social-construct – a projection of my own values based on an *idea* I inherited from my forebears.[105] Perhaps that's fine up to a point. But I now think the core idea of wilderness is out of date. Preserving pristine nature isn't a helpful ethos anymore – if it ever was. Like it or not, we live in a world fashioned by humans. That gives us a

---

99   Kolbert, 2021, pp. 6–8.
100   Parker, 2020.
101   McKibben, 1989, p. 48.
102   McKibben, 1989.
103   Donner, 2007, p. 232.
104   Preston, 2012, p. 191.
105   Callicott, 1991; Cronon, 1995.

responsibility to manage that world well – *even if* that violates our romantic traditions.[106]

Hope: That strikes me as hubris, plain and simple.

Carlos: I agree that *pristine* nature is a problematic concept. I've seen it used to justify removing indigenous populations off the land too many times, especially from national parks.[107] Human beings are seen as intruders or unnatural, and so we remove them. This attitude ignores the fact that humans have always lived as part of nature, and in doing that their presence has affected how the Earth has changed.

Richard: Yes! The alleged "wildernesses" of our national parks have been shaped by indigenous peoples over thousands of years, and many can only be maintained now with constant management. [108]

Hope: There are real issues here. Nevertheless, geoengineers don't compromise only the *purity* of the natural world: they aim to eliminate its *independence* entirely.

Richard: Why would that be a problem?

Marcia: For one, I think there is wisdom in tradition: once, many of our ancestors respected the independence of other creatures and systems. In my home, the Xerente believe that rivers and forests and animals hold an independent soul.[109] Any attempt to catch these souls in our fingers, using them for our ends, violates the obligation to respect these entities, in part because it attempts to obliterate their independence.

Hope: I remember that Chief Standing Bear, a Lakota philosopher, once said that "the great distinction between the faith of the Indian and the white man [is that] Indian faith

106   Marris, 2011, pp. 170–171. See also Hobbs, *et al.*, 2011.
107   Guha, 1989.
108   For a collection of readings on the conceptual and practical problems with the idea of wilderness, as well as for myriad attempts to defend it, see Callicott & Nelson, 1998.
109   Inoue, 2018, p. 37.

sought the harmony of man with his surroundings, while the other sought the dominance of surroundings."[110] He thought that independent nature has value of its own, and to crush it is the essence of narcissism.[111] What is SSI if not further domination of nature?

Richard: I think you're exaggerating the degree to which geoengineering bends the natural world to the human will. Even the simplest technologies exceed our intentions to some extent.[112] Metal rusts; buildings collapse; cars stall. To geoengineer isn't to obliterate the independence of nature. On the contrary, I hope it will allow species and ecosystems to flourish in a climate more suitable for them.

Hope: Maybe you're right, but isn't that a double-edged sword?[113]

Richard: How?

Hope: I agree that we don't have the technology right now to fully control the ecological world. So, there will be plenty of independent nature under SSI. But isn't this the same worry that we've raised time and again? We don't know and probably can't fully predict the effects of SSI.

Marcia: I believe massive geoengineering interventions are arrogant either way. If they work, they express the human narcissism that the natural world should operate exactly how we desire. If they fail, they demonstrate that we lacked the proper humility to recognize the limits of our understanding.[114]

Richard: What do you mean "lacking proper humility"?

Joseph: For one thing, it's playing god. That will invite catastrophe.

---

110 Standing Bear, 1933, p. 205.
111 Lee, 1999, pp. 201–203.
112 Vogel, 2015, p. 112.
113 Preston, 2012, p. 194-197.
114 Obst, in preparation, 2.

Richard: That's just another metaphor. Or do you literally think that stratospheric sulfate injection is wrong because human beings would take the place of God?

Joseph: (*smiling*) Perhaps I should not have used that expression. I can see it might be too theological for some.

Richard: Anyway, perhaps we should *embrace* our new role.[115] Don't you think there is something *exciting* about humanity reaching out, taking control and molding the Earth? Maybe it's our destiny.

Kim nods; Marcia gasps, staring back at Richard in shock.

Hope: (*intervening*) Maybe part of what bothers Joseph and Marcia is that climate engineering would, in effect, make humanity directly responsible for the global climate. I don't think any person or government – indeed any institution we have – is prepared to take on that responsibility.

Kim: What do you mean "prepared"?

Hope: I mean equipped – socially, politically, and ethically. No national government, for example, has the political legitimacy to take on managing the global climate on behalf of humanity as a whole. And no one is ready to take on the ethical responsibilities that likely emerge, such as for protecting those vulnerable to negative impacts of geoengineering, or compensating those who actually suffer bad effects.

Marcia: Ready or not, I don't think we should be going in this direction. Some forms of geoengineering, like SSI, aim at taking control of the *basic physical structure* of the planet. To me, that dramatically extends humanity's reach, and in particular the reach of the powerful who will exercise the control.[116]

115   Lynas, 2011.
116   This discussion draws on Gardiner 2016b, pp. 125–127. This reality, while probably a way off, may not be as distant as one would assume. Christopher Preston argues that humanity stands on the cusp of a *Synthetic Age* where human beings will deliberately shape innumerable operations once left to nature. See Preston, 2018.

Richard: So?

Susan: It *politicizes* the climate in a deep way.

Joseph: Those doing it would owe the others a lot, as a matter of basic justice.

Peter: Why? What if taking control of the basic physical structure protects people? What if it promotes justice and enhances welfare? What would be wrong with that?

Joseph: I doubt that it *would* promote justice and enhance welfare. But even if it might, there may be good reasons to resist. Suppose some super-duper neuro-scientists said they could promote justice and enhance welfare if we let them take control of our brains – our very thoughts. Would that seal the deal for you?[117]

Amina: Yuck! No way. I wouldn't want someone inserting thoughts in my head, even for the sake of some greater good. I'd be losing myself.

Peter looks thoughtful.

Marcia: I'm beginning to realize how fundamental our disagreement is. To my mind, a *truly radical* solution to climate change – one that gets to the roots of the problem – requires a deep shift in humanity's relationship with the environment. Human domination over the environment is morally tainted in the best of circumstances. If there is to be an answer to the problem of control, it cannot be more control.

Joseph: I am also concerned. As I see it, climate injustice is, at root, a social phenomenon: a product of unconstrained capitalistic consumption, humanity's alienation from the natural world, and environmental racism. I also believe white supremacy is at the heart of it.

Richard: You've really lost me.

Joseph: In my view, environmental racism is a problem of *racist environments*: it's symptomatic of more fundamental

117 Gardiner 2016b, pp. 127–128.

problems with how humans experience and shape the world.[118] Black and indigenous people today are seen as little more than trash, if they are seen at all.[119] Their interests are ignored.[120] My point is that masking the effects of climate change through geoengineering doesn't change these racist environments: it deepens them. The more control, the worse it gets. So, we can expect SSI to continue the pattern, imposing more unequal environmental harms on the most marginalized.

Richard: With all due respect, if the biggest driver of environmental harm is race, how do you think communities of color will fare in a world of 2 or 3 degrees warming? The marginalized will have less access to air conditioning to endure the heat waves, less access to water to endure the droughts, less access to diverse sources of food to endure agricultural declines, and less money to relocate as shorelines creep inward and wildfires rage.[121] And what about 5 degrees or more? That would be infinitely worse – a complete catastrophe. Technological interventions could prevent all this. That is why I think wealthy countries have *moral obligations* to geoengineer on behalf of marginalized people worldwide.[122]

Joseph: By saying so, you only make me more certain that geoengineering would not benefit my people. Rather than respecting our right to self-determination, you paternalistically suggest *you* will save us – from yourselves.

Richard: (*angrily*) Ultimately, someone will use this technology. If not the United States, another country. We at least are trying to include you in the conversation. Yet you accuse us of paternalism!

---

118 Opperman, 2019, p. 59.
119 Mills, 2001, p. 74.
120 Bullard, 2001, p. 165.
121 Horton & Keith, 2016, p. 79.
122 Horton & Keith, 2016, p. 80.

Joseph: You may be responding to my words, but you haven't *heard* a single thing I've said.

Abrupt silence.

Hope: (*awkwardly*) Perhaps we should take a break, give ourselves a chance to reflect ...

Richard: Agreed. That's enough for today. I'm sure we'll need to revisit these issues as time passes.

Kim: Thank you for organizing this meeting; we found it most instructive.

Richard: Thanks for coming. (*rising*) Looking ahead, I'm thinking a smaller group may work better, one made up of those who are motivated and share common ground. I'll be in contact with *some of you* soon.

He leaves without a farewell to Hope; others start to drift out. Marcia and Amina walk over.

Amina: Hope, we wanted to thank you for your efforts. May we stay in touch?

Hope: Please do.

Marcia: I fear I may have seen the beginning of the end of the world in this room today. But I will not succumb to that fear. We must be strong. A path may reveal itself. I still believe that our Mother Earth will not abandon us; she will find a way to give us what we need.

Dialogue 6   The Future

Sometime in the 2060s

Part 1: Discounting the Future

Another sweltering evening in Washington, D.C. Hope sits on the porch, sweat trickling down her wrinkled skin. Humid and hot as it is, she is determined to breathe the fresh air. She glances at her phone idly: just past noon, August 14th. Without a word, the screen dissolves into a weather report – heat index 103.8°F. When she started in Congress, there might have been a week of such days all year, but now they define the late summer.[1] *This year is on track to be the hottest year on record, as is the decade. The Earth is now around 2°C warmer than preindustrial averages, with models predicting more warming locked-in.*[2] *The ongoing warming has increased desertification,*[3] *pushed an eighth of all species to extinction,*[4] *and nourished run-away mosquito populations, spreading disease around the globe.*[5]

Hope takes a sip of her iced tea and closes her eyes with a sigh, appreciating the breeze from her feeble fan. She received it as a present from her husband almost thirty years ago, when she

1 Fenston, 2019.
2 In AR6, very high, high, and moderate emissions scenarios all project 2 degrees warming by 2060, with low predicting about 1.8 (IPCC, 2021b, p. 29).
3 Burrell, *et al.*, 2020.
4 Urban, 2015.
5 Kraemer, *et al.*, 2019.

DOI: 10.4324/9781003123408-6

first won her spot in Congress. She'd campaigned on a platform of addressing global warming, so Mark thought a fan would be funny. She hadn't appreciated his humor at the time. A ghost of a smile flitters across her face, and she savors the pleasant moment.

That is important to do these days. As ever, missed temperature targets and empty political promises abound. Still, so does fierce resistance. Back in the '20s, people from diverse backgrounds had formed the Climate Justice Consortium (CJC), with the goal of defending the interests of the vulnerable and marginalized worldwide. By 2040, they had amended their mission statement to include a pledge to defend the interests of the Earth. The CJC now counted ecosystems and nonhuman life as entities to be cherished and protected for their own sake. Hope had been asked to propose this historic amendment for formal ratification. It had been the honor of her life.

The very same year, of course, the first bill to authorize emergency deployment of stratospheric sulfate injection[6] found its way to Congress. It had failed. But as things worsen, calls for SRM solutions have grown more popular each year, their risks seen as more and more acceptable by many. Although no amount of adaptation can resurrect the economy or the dead, many are increasingly tired of weathering the storm at such great loss. They want to "take back control"; or, at least to try. Just last year, Hope worked around the clock to fight another climate engineering bill. Those incredible young people she worked with were as energetic and clear-eyed as young people had ever been. She found their actions inspiring, even in desperate times.

Something clatters from behind Hope, and she turns to see Caitlyn, her 16-year-old granddaughter, squeeze through the screen door in a summer dress the color of the lemonade she holds in her hand. The cup's plastic sweats as soon its cold surface hits the air.

6 See Dialogue 5.

Caitlyn: Hi Grams.
Hope: Hello, Caitlyn. How's my sunflower?
Caitlyn: I'm okay. How are you? I was watching from the window. You looked sad.
Hope: Just thinking about work, honey.
Caitlyn: The warming again?

Hope nods, somewhat reluctantly.

Hope: Did you know that days like these were once rare? I used to look forward to them: perfect pool days. Now, with pools closed and the heat so common, they have lost their charm.
Caitlyn: (*shrugs*) They're normal to me. I've never really known anything else.[7] (*Worry flits across Hope's face*) It's okay, Grandma; the heat itself doesn't bother me much, compared to some of the other impacts. I especially feel for all those people without air conditioning, a home to sleep in, or food to eat. I know how unequal the effects are.
Hope: I've tried to help my entire life, but it seems I didn't do enough. I fear for your future .... (*Caitlyn's eyes widen*) ... I'm sorry, I don't mean to scare you. You're right. We are lucky to live where we do. Your parents and I will protect you.
Caitlyn: I know, Grams.

They sit in quiet, disturbed only by the distant wind turbines humming along with the sounds of summer. Those turbines were some of the very first built, standing antiques still spinning.

---

7 Insights in environmental psychology reveal that humans assess environmental change relative to the baseline environment they knew as children. While perhaps unsurprising, this psychological feature results in an easily overlooked – yet pernicious – phenomenon: new generations lack the ability to fully appreciate the ways their environment has deteriorated relative to a far-gone baseline they never knew. Peter Kahn calls this "environmental generational amnesia" (Kahn, 2002).

Caitlyn: I try not to fear, but sometimes I can't sleep thinking about all the suffering.[8] How did things get so bad, Grandma? Why didn't people act sooner?

Barriers to Climate Action

Hope: I ask myself that question every day. Have you learned about climate change in school?

Caitlyn: Of course. Almost every subject has a unit on climate change now, but it's all about the facts. They never really explain why serious action took so long, or why it continues to be inadequate. Why you were so alone in your efforts.

Hope: I had more help than you know, child. But you are right; I met much resistance. I continue to. And if you can believe it, I was really in the second wave of climate activists. People knew about the greenhouse effect before I was born, and environmentalists clamored for prevention[9] and mitigation just as long, but still little was done.

Caitlyn: Why?

Hope: It's complicated. If you ask an economist, they might tell you that climate change was just an incredibly costly market failure. If you ask a political scientist, they might blame a lack of adequate institutions and perhaps a fear – leftover from the "Red Scare" – of top-down, socialist climate solutions. If you ask an ecologist, they might

---

8 Recent research in psychology testifies that climate anxiety has become common among young people today, with 60% saying they were either very or extremely worried about climate change. See Marks, *et al.*, 2021.

9 Referring to the late 1980s and early 1990s, Dale Jamieson reports: "In those halcyon days, 'prevention' referred to the possibility of preventing climate change. We now know that it was not possible to prevent climate change even then, much less now" (Jamieson, 2014, pp. 201–202). Of course, that is not to say that *dangerous* climate change of more than 1.5 or 2 degrees could not have been prevented back then.

say that most people didn't understand the Earth well enough, early enough. (*Hope smiles slightly at Caitlyn's look of confusion*) Still, I've always thought of myself as a philosopher. So, I guess it's not surprising that I believe that, although my generation did fail economically, politically, and ecologically, our largest failure was *ethical*.

Caitlyn: Your generation was immoral?

Hope: (*hesitantly*) Yes.

Caitlyn: Why?

Hope: I can only answer that question from my perspective, and some would disagree with me even now. But do you want to hear my thoughts?

Caitlyn nods eagerly, and Hope laughs.

Hope: I hope you have plenty of lemonade. This may take a while. To begin, climate change wasn't like any other moral problem. Even when I was your age, I knew of global warming, and learned about the greenhouse effect in science class. But it wasn't until college, when I took more advanced classes, that I realized just how many barriers stood in the way of real climate action.

Caitlyn: What barriers? All your generation had to do was stop burning fossil fuels.

Hope: In a way, it *was* that simple, and it's refreshing to hear you point that out. And, yet, it was also horribly complicated. One barrier was the powerful economic incentives some had to keep burning fossil fuels in the short term.

Caitlyn: You mean like the old fossil fuel companies? Learning about them filled me with such rage. Amazingly, the CEOs of the companies most responsible could fit inside three buses.[10]

Hope: (*sighing*) Some of the corporations, who practically ran the world when I was a young girl, have now become the

10 The climate scientist Katharine Hayhoe offered this striking visual in an interview with Jimmy Kimmel on September 22, 2021 (see Hayhoe, 2021).

villains in our history books. And deservingly so. The way they fanned climate denial for profit was grotesque. Yet it's also important to remember that for a long time they were allowed to do what they did. We needed early, strong regulation of fossil fuels urgently, but it took far too long to happen.

Caitlyn: Why was that?

Hope: Well, answering that leads to the second major barrier: the problem was truly global. Climate change affects everyone, and the entire world needed to come together to fix it. Yet national governments frequently justified their lack of action on fossil fuels by pointing to other countries' inaction. It created a vicious circle, an ongoing cycle of finger pointing. There were no international institutions capable of breaking this cycle, even though we *needed* such institutions direly. As you know, this problem continues today in many respects.

Caitlyn: Why haven't we created them? We've had lots of time.

Hope: In my view, the real reason for resistance is one nobody wants to admit: effective global institutions would require national governments to relinquish some of their power. They don't want that; and they use their power to resist it.

Caitlyn: I've heard people say that the United Nations should have more power. Would that work?

Hope: I'm not sure the UN is the right venue. It is structured so as to make it closely tied to national governments, and it tends to be dominated by a few powerful countries. But I have long thought that we need better institutions to deal with these kinds of global, intergenerational problems.

She smiles to herself, remembering Eliza's proposal, all those years ago.

Caitlyn: Don't you worry that global institutions, if granted the power needed to hold national governments accountable, would necessarily become despotic and tyrannical?

Hope: Yes! That is a real concern. Nevertheless, I think it's often overblown. Virtually all institutions carry this risk, including many we already have. Still, the risk can be contained. Institutions can be designed to minimize it, and it helps if there's a strong public culture where people are on their guard. One can be skeptical, of course, but I don't see why worries about despotism are often treated as obviously decisive when it comes to global problems, but almost nowhere else.

Caitlyn looks thoughtful.

Hope: Another barrier is visibility. This was especially true earlier in the twenty-first century when governments couldn't *see* the adverse effects of their fossil-fuel addiction, spread as they were across space and – more importantly – time.

Caitlyn: More importantly?

Hope: Yes. Climate change is a deferred phenomenon, and this was a significant barrier to action. For a while, climate action was primarily a moral duty we had to the future, to the young, and future generations. Many of the government officials we needed to take action would never experience the worst consequences of their failures, nor would the majority of the people voting for them.

Caitlyn: I guess I don't understand why the time-lag should matter. If someone planted a bomb underneath a city set to go off 100 years in the future, we all know that would be morally wrong!

Hope: Very true.

Caitlyn: Besides, people care about their children and grandchildren, right? Why didn't they care about climate change?

Hope: Of course, parents and grandparents care! (*reassuringly*) Even so, it was widespread, and often considered justified, to weigh the interests of current people here and now more heavily than people in the future.

Caitlyn: How could *that* be justified? Sounds like you're trying to have it both ways.

Economics and the Social Discount Rate

Hope: Here's one way to justify it. When making policy decisions, don't you think it's helpful to know the pros and cons of those policies?

Caitlyn: Of course!

Hope: Well, people, especially politicians, often turn to economists to get this information. They ask for a cost–benefit analysis (or CBA).

Caitlyn: Wait. Is an economist's cost–benefit analysis the same as a pros-and-cons analysis? They sound different to me.

Hope: Great observation! They are different.[11] Identifying "pros and cons" is not controversial. It leaves a lot open – like what counts as a "pro" and what as a "con," and how we are supposed to compare them. Thus, as a method it can be fleshed out in numerous ways.

Caitlyn: I get that. Sometimes I make a list of pros and cons when I'm trying to make a difficult choice – like whether to stay home and do my homework or hang out with my friends. But what about cost–benefit analysis?

Hope: That goes beyond simply making a list of pros-and-cons. In the first step, the basic idea is that you can specify the costs and the benefits of a project or policy, and then add them all up together. If a project is a net benefit, then it passes the test and should be done; if it doesn't it should not be done.[12] I call this step, 'net-benefit analysis'.[13] It assumes that you can compare costs and benefits in a way that makes sense of the adding together.

11 Our discussion of CBA draws on Gardiner, 2016b, pp. 76–82, and Gardiner, 2011a, chapter 8.
12 Frank, 2005.
13 Gardiner, 2016b, p. 77.

Caitlyn: How is that comparison done?

Hope: That's the second step. In principle, you might evaluate what counts as a cost and a benefit in numerous ways. But in practice, people generally employ the standard economist's method, which is to use the market prices of goods. I call that "standard market-based CBA."

Caitlyn: Hang on. Doesn't that presuppose that we can measure the value of everything in terms of market prices? I'm not sure I buy that. (*smirks at the pun*) I don't think the value of many of the things I care about is captured by what it would cost to *buy* them. My friends, for example. Also, I don't think I'd want a friend who could be "bought"!

Hope: Fair enough. But do you want to know more about how the method works? Half a century ago, two prominent economists stepped up to this task: William Nordhaus and Nicholas Stern. Both were highly influential. Nordhaus set out a pretty standard economic approach; but Stern thought his conventional assumptions were inadequate for a problem like climate change. Ultimately, Nordhaus' economic model recommended much less aggressive climate action than Stern's.[14]

Caitlyn: Why was Nordhaus' model less aggressive?

Hope: The biggest factor was that they valued the future differently.[15] Economists typically count benefits and costs that occur further in the future as less valuable than those that arise sooner.

Caitlyn: You mean like a dollar earned today is worth more than a dollar in twenty years?

---

14   Nordhaus, 2007; Stern, 2006. For an explanation of Stern's central concerns with Nordhaus' approach, see Stern, 2008.

15   Another issue is that there are significant differences in methodology between the descriptive and prescriptive approaches which mean that it is not clear that both are trying to answer exactly the same questions (Arrow 1996; Kelleher 2017). Still, both claim to be policy-relevant, and that fuels much of the interest in them and the debate.

Hope: Not exactly. That might be due to inflation, the general increase in the price of goods and services over time that reduces the purchasing power of a given amount of money. You know, like I used to able to buy a candy bar for a dollar, but not now.

Caitlyn: Really – a dollar? You must be old! But discounting is not about that?

Hope: Nope. It is about the real value of money at different times, even after we've considered inflation.

Caitlyn: Go on.

Hope: Economists typically employ a positive "social discount rate" (SDR). They value the future less than now, and at a uniform rate.[16] The higher the number, the less valuable benefits later in time are compared to those closer in time. Almost every economist uses an SDR, but they often disagree about what the rate should be. Usually, they choose somewhere between 1% and 10% per year, depending on the context. Rates of 2%–5% are common in policy analysis. Obviously, the choice of SDR matters a great deal if you are discounting over long periods, such as decades or centuries. For example, think about how many benefits in 100 years it would take to equal one benefit now: with an SDR of 1%, you'd need 2.7 benefits; at 3% it would be 19.2; at 5% it becomes 131.5; and at 10% around 13,781. So, it makes a *huge* difference which number you pick.

---

16 Positive rates imply that future costs and benefits are worth less than those that arise today. Some define discounting so that the fact that the future is worth less is part of the definition (e.g., Toman 2001, p. 267: "A method used by economists to determine the dollar value today of costs and benefits in the future. Future monetary values are weighted by a value <1, or 'discounted'"). We are sympathetic, since this makes sense of the term *dis*counting. However, it has now become common to define the SDR so as to allow that in principle it could have a negative rate (e.g., –1%). This would mean the future would be valued more highly than the present. That being said, we note that in practice – in actual CBA – negative rates are extremely rare.

Caitlyn: I see the basic idea. Can you give me a concrete example?

Hope: Sure. I once heard an economist say that if you take the value of the whole global economy in 200 years, and then apply a discount standard rate of 5%, the value today would be just a few hundred thousand dollars. That implies that if you were trying to decide how much it is worth investing *now* in preventing the destruction of the Earth *in 200 years*, a simple CBA would tell you the answer is no more than you would currently be willing to invest in an apartment.[17]

Caitlyn: That's ridiculous!

Hope: Maybe. It's certainly disconcerting.

Caitlyn: (*impishly*) So, how did these old economist guys do their discounting?

Hope: Nordhaus discounted at a rate of 5.5% per year; Stern at the much lower rate of 1.4%.[18] This is the central place where they disagreed.

Caitlyn: Okay – given the compounding effects, that's obviously a massive difference. I assume they have their reasons. Still, so far I haven't heard any *explanations* for the practice of discounting that might explain the difference. Why do economists discount the future at all? Surely a benefit is a benefit, whether you have the benefit now or in the future.

Hope: That's an excellent question, with a bit of a complicated answer. Discounting is primarily a *practice* – something economists *do*.[19] It turns out that there is not one, single reason why. Instead, there are multiple justifications in play, and no general agreement about which is the most important, which can create confusion. Worse,

---

17 Chichilnisky, 1996, p. 235. The calculation assumes we should value the Earth on the basis of measuring the economic value of foregone output.

18 These are simplifications, drawn from Nordhaus, 2007. See footnote 20 below.

19 Gardiner, 2016b, p. 81.

sometimes the various justifications pull in different directions.

Caitlyn: Okay. Can you tell me about the various reasons?[20]

Hope: Sure. One prominent reason to use a discount rate is that people *just do* prefer having a benefit now rather than in twenty years. This is called *pure time preference*.

Caitlyn: I think I understand that. I'd rather have an ice cream cone now than in a week. But I think that's kind of short-sighted of me.

Hope: Right! (*beaming with pride*) Perhaps, even when it comes to self-interest, pure time preference shouldn't be indulged. Wouldn't it be lovely next week to remember you saved an ice cream cone as a treat for your future self?

Caitlyn: Yeah. (*smiles*) That would be nice.

Hope: More importantly, I'd emphasize that when we're talking about the economics of climate change many of the future costs and benefits we're talking about will go to *other* people. Given that, *pure time preference* might become not only short-sighted but also deeply *unethical*.[21] It doesn't seem right to count impacts on others in the future as less important *just* because *you and I* have a short-term bias. For example, I'm thinking that the old economists shouldn't have discounted, say, the negative impacts of losing most of Miami and Bangladesh to sea level rise forty years later based on pure time preference.

Caitlyn: That's what I was thinking.

Hope: Some economists and most ethicists agree. But not all. Nordhaus, for instance, does include a substantial *pure time preference* in his discount rate; Stern does not. So, there appears to be an ethical disagreement there, lurking inside the economics.

---

20 In the following discussion of possible justifications, we often draw from Parfit 1983b. See annotated bibliography.

21 Parfit, 1983b.

Caitlyn: Interesting! What about the other reasons for discounting?

Hope: Well, some argue that benefits in the future should be discounted because they are *less likely* to occur. Take the ice cream cone example. The power may go out, the freezer turns off, and the ice cream cone melts. If that happens, you won't be able to enjoy it in a week.

Caitlyn: That would be awful!

Hope: (*laughing*) Yes. Still, the philosophical point is that the benefit of eating the ice cream now is basically guaranteed, but the benefit in the future is not.

Caitlyn: I can see that.

Hope: Still, we should be careful. Notice that it is not really the future ice cream cone itself that is less valuable. Instead, we're discounting for the probability that the cone will be available. That's a significant difference. For instance, bringing it back to climate, it seems justifiable to value preventing deaths from an actual, impending hurricane greater than preventing deaths from a *possible* hurricane in the future because the danger associated with the imminent storm is much more likely. But that's entirely different from saying that saving current people's lives is more important than saving the lives of future people

Caitlyn: Yeah, I understand why probability matters. (*Slowly, face wrenched in thought*) To make this clear, maybe we should call that discount rate a *probabilistic discount rate* instead.

Hope: That's a good idea. We should also emphasize that probability does not always *go down* with time. For instance, in my youth some Americans petitioned the government to spend more resources on disease response, knowing that a pandemic would eventually break out, and we would need to be ready. Even though the likelihood of a pandemic happening in any given year was low, there was an extremely high likelihood it would happen *eventually*. And, sadly, those worry-warts were right. In 2020, as I'm sure you've learned, COVID-19 broke out, and hundreds of thousands of American lives were lost.

Caitlyn: So, you're saying some things grow *more likely* as time goes on?

Hope: Yes.

Caitlyn: If that's true, probability can't always justify a conventional discount rate over time, can it? (*Hope shakes her head*) Are there any other justifications, then?

Hope: Another justification is the argument from opportunity cost. Do you know what an opportunity cost is?

Caitlyn: Nope.

Hope: Let me explain. Let's go back to the ice cream cone. Imagine you have one cone, and you want to store it in the freezer for later. The benefit of doing so is one cone to eat in the future, right? (*Caitlyn nods*) Okay; so, you are about to put the cone in the freezer, when you remember there are four other cones outside on the porch. If you store the first before collecting them, you're sure the others will melt. Following?

Caitlyn: Yeah. I should collect the other ice cream cones first, so I have five cones to enjoy later.

Hope: That would be the smart thing to do, wouldn't it? The four ice cream cones melting in the sun is the *opportunity cost* of putting the first ice cream cone away first. Of course, you could say that putting the first ice cream cone away has positive value: one cone to eat later. But that's a bit misleading because you have the *opportunity* to have five cones instead! In other words, factoring in the opportunity cost, you recognize that the choice to store the first cone has disvalue.

Caitlyn: I get it now. How does this relate to justifying a social discount rate?

Hope: Suppose we think in terms of standard investments. One reason a dollar right now is often worth more than a dollar delayed is that you have the *opportunity* to invest the current dollar. In the time it would take to receive the delayed dollar, that current dollar would steadily gain in value as the returns on investment come in. So, by the time the current dollar arrived at the same place in time as the delayed dollar, the current dollar would be worth more than one dollar. See? In this case, delaying some benefits involves an opportunity cost.

Caitlyn: So, in the climate case, we're talking about investing in economic growth, right?

Hope: Generally. For instance, both Nordhaus and Stern discount for economic growth. It accounted for 2% in Nordhaus' overall SDR of 5.5%, and 1.3% in Stern's SDR of 1.4%.[22] So, they differ, but not by that much and this doesn't drive most of the overall difference between them. Interestingly, most economists when I was younger simply assumed that the economy would keep growing through all the changes. Their core disagreements were really about other things.

Caitlyn: Hang on. Not all delays in consumption result in more benefits in the future, right? If I eat my ice cream now,

---

22 Discounting is usually understood in terms of the Ramsey equation, developed by the mathematician and philosopher, Frank Ramsey. This equation states that $\rho = \delta + \eta\gamma$, where ρ is the discount rate, δ is the pure time preference, γ is growth, and η is a complex variable called the consumption elasticity, sometimes thought to cover risk, inequality aversion, and other factors. Nordhaus' 5.5% discount rate (for the first century) has a pure time preference of 1.5, 2 for growth, and 2 for the consumption elasticity. Stern's calculations are less explicit due to a different background methodology. However, Nordhaus provides a useful benchmark by estimating them at 1.3% for growth, 1 for η, and 0.1% for extinction, which Stern attaches to δ. Some suggest Stern's discount rate (for the first century) is higher – e.g., 2.1% for the twenty-first century – and worry that even higher returns could be earned on investments, especially in developing countries (e.g., cf. Beckerman & Hepburn, 2007). For discussion, see Gardiner 2011, Mintz-Woo, 2021.

rather than saving it for next week, I'm not missing out on any opportunity, am I? It would just sit in the fridge otherwise. It wouldn't make more ice cream!

Hope: Brilliant, Caitlyn. Another way of putting your point is that, yet again, *time* and opportunity costs do not always correspond.

Caitlyn: So, if the extra benefits in the future won't actually arise, you can't justify doing discounting *as if* they would, right?

Hope: Yes. The argument from opportunity costs fails, or at least becomes irrelevant, if you consume the resources now rather than investing them.[23]

Caitlyn: What if you invest the money poorly? If the so-called opportunities result in losses, rather than benefits? What if you know – or should have known – you're making bad investments, or ones that are really high risk?

Hope: I can guess where you're heading. You're thinking we shouldn't just *assume* continuous economic growth in the future.

Caitlyn: Yeah. To me, those old economists sound way too complacent. They didn't take seriously enough the threat that climate damages might cause the economy to contract. Look around. I don't feel like I've been growing up in a period of growth.

Hope: I wonder. Perhaps technically …

Caitlyn: Pah! Give it up, Grams. Even if some economist could persuade me that the economy has been growing in some *technical* sense, that would just make me think that what they're trying to capture is not what matters for *genuine prosperity*.

Hope: Hmm. That reminds me, some argue for a social discount rate because they assume their successors will be better

23 E.g. "If we will not pay compensation … it becomes an <u>irrelevant fact</u> that it would have been cheaper to ensure now that we could have paid compensation" (Cowen & Parfit, 1992, pp. 135–153)."

off. For a long time, globally, quality of life trended up as technology improved and infrastructure grew.

Caitlyn: But that's what I'm objecting to!!! The future is not always better off. I don't mean to sound spoiled, I know I don't have it that bad personally, given everything. But I still think your generation's world without continuous fire, flooding, and famine was far preferable to the world I'm growing up in.

Hope: I agree, sunflower. I don't think it is responsible simply to assume that our successors will always be better off.

Caitlyn: (*grumpily*) Yeah – discounting their interests at 5% per year under these conditions might be a pretty reliable way of ensuring that they're not. (*collecting herself*) Anyway, why are we talking only about investments, resources, and economic growth? I'm worried that standard CBA misses too much, including the real value of many things I love.

Hope: That reminds me of a criticism some have of discounting, that it doesn't seem to work for all benefits. For example, consider the decision to build an airport on a stretch of beautiful countryside. If we do not build the airport, the countryside will be enjoyed every year going forward. However, if we applied a social discount rate to its beauty, that beauty would be valued less in twenty years time than it is now. That doesn't seem to make sense. The benefits of its beauty experienced next year and every year after cannot be reinvested like a dollar could. So, there is no reason to discount for opportunity cost for those kinds of benefits.

Caitlyn: Some benefits of the Earth are like that, I think.

Hope: It's a worry; though, to be honest, it's common for those practicing CBA simply to *exclude* benefits like natural beauty from their calculations altogether. So, often the problem of discounting them doesn't even come up.

Caitlyn: That doesn't make things better! Doesn't exclusion make it more likely that their value is simply ignored in policy?

Hope: You may be right.

Caitlyn looks even more unhappy.

Hope: What's the matter?

Caitlyn: These various reasons for discounting just don't seem
         very strong. Yet you say almost all economists did it and
         many still do. I can't help but feel like they're just being
         selfish and wanting to advantage their own generation.

Hope: Maybe a few are guilty of that; but I'm sure the over-
         whelming majority are not. Let me mention two final
         reasons to see if you change your mind at all.

Caitlyn: Okay.

Hope: Another reason why economists discount is that, if
         they didn't, it seems too easy for benefits to the future
         to outweigh benefits today. The number of people who
         will exist in the future probably far exceeds the number
         alive today. Therefore, if their interests are weighed the
         same as the interests of current people, it seems likely
         that future people will usually win out – perhaps almost
         always! That seems overly demanding. Isn't it unrea-
         sonable – and unethical – to expect the current genera-
         tion to spend their life in the service of the future? That
         would be an excessive sacrifice. To avoid it, we might be
         tempted to discount.[24]

Caitlyn: I get it. Still, does discounting by itself help that much?

Hope: Why do you say that?

Caitlyn: Surely it depends on how big the benefits in the future
         are. If they're *really huge*, then the current generation
         might still have to sacrifice a lot.

Hope: Hmm. I hadn't thought of that.

---

24 For instance, the economist David Pearce raises the following objection to
zero rates of discount: "everything would be transferred to the future ... pure
equality of treatment for generations would ... imply a policy of total current
sacrifice" (Pearce, 1993, p. 58).

Caitlyn: Anyway, what about this excessive sacrifice thing? I think the moral thing to do is to respect the interests of each generation, and that requires not making one generation a mere servant to the interests of other generations.

Hope: I see your point. Kant famously argued that it was wrong to treat another person as a mere means to our ends. He saw that principle as fundamental.[25]

Caitlyn: That sounds good to me.

Hope: I wonder if what we really believe is that no generation should be expected to make certain kinds of *extreme* sacrifices for the sake of the future.[26]

Caitlyn: Sounds right to me.

Hope: Since we're talking about Kant, there's another issue related to the extreme sacrifice worry. CBA is focused on overall economic welfare. It usually doesn't concern itself with how that welfare is distributed, but concentrates on maximizing the total – the size of the economic pie, if you like. In doing that, it is like an approach to morality known as utilitarianism: the view that the right thing is what maximizes welfare.[27] But utilitarianism is notoriously vulnerable to the criticism that it too readily sacrifices individuals for the sake of benefitting the group. For example, it might require punishing an innocent person if that would appease an angry mob. Many people think that would clearly be wrong, and so reject utilitarianism.[28]

---

25   Kant's moral system has one ultimate principle, but he expresses it in three different formulations. One formulation – the formula of humanity – states that "we should never act in such a way that we treat humanity, whether in ourselves or in others, as a means only but always as an end in itself" (Johnson & Cureton, 2016).

26   Questions also arise about whether the current generation should have to sacrifice anything if what's at stake is just making the future even better off than we are now. For criticism of this idea, see Gardiner, 2017b.

27   Much climate economics is explicitly carried out within the framework of utilitarianism.

28   Others argue that CBA is a very simplistic version of only one variety of utilitarianism, namely act-utilitarianism, and that other forms of utilitarianism – such as rule utilitarianism – are better (cf. Hooker, 2015).

Kant, for example, thinks we shouldn't treat people as mere means, even if it maximizes welfare to do so.

Caitlyn: Interesting. How does that dispute play out here?

Hope: Well, it suggests that some of our problems with CBA might ultimately be with utilitarianism. CBA claims that a project is worth doing if its benefits exceed its costs. But that ignores questions of justice, freedom, rights, and personal integrity. Many people think those values are central, and at least as important as economic values. In short, it would be a problem if CBA simply assumes a controversial theory of ethics – namely, utilitarianism – and a crude version at that.[29]

Caitlyn: That's interesting. I'd like to learn more about these theories. But for now, what's the last justification for discounting you wanted to raise?

Hope: Oh, right. The idea there is that there is always the possibility that an unforeseen cataclysmic event could wipe out humanity.[30] In such a case, future welfare would not exist, and therefore would have no value. This, of course, is *really* a probability discount rate, which we saw before.

Caitlyn looks profoundly dissatisfied.

Caitlyn: Let me get this straight. Are you saying that it's justifiable for a generation to invest less in protecting the future *from a threat they created*, because there's a risk humanity might go extinct anyway, in which case the investment wouldn't pay off and the money would be wasted? That seems really weak to me.

---

29 The relationship between conventional CBA and utilitarianism is controversial and contested. For relevant discussion, see Gardiner, 2016b, pp. 73–76 ; for a nonutilitarian defense of CBA, see Schmidtz, 2001.

30 Kian Mintz-Woo argues that the extinction risk is not privileged, but is instead part of a class of events that prevent our initial choice from changing the value of the subsequent outcome (Mintz-Woo, 2019).

Hope: I see your point. It sounds like a severe abdication of responsibility.

Caitlyn: Also, I think the whole thing might be morally corrupt. Notice that the consequence of your generation's alleged "fear of human extinction," and of wasting resources on trying to protect us, is that you get to keep more of your resources, to spend on things like luxury goods.

Hope: Ouch.

Caitlyn: Frankly, a lot of these arguments for discounting seem way too convenient and self-serving.

Hope: I share your frustration with social discount rates, Caitlyn. Perhaps the nicest thing that can be said for them is that sometimes a social discount rate can roughly correspond with other factors – like probability, opportunity cost, and unreasonable burden – that *are* practically and morally relevant.

Caitlyn: The whole thing makes me angry. I just can't help but feel like many of these justifications were just convenient ways to rationalize not caring about the future. If we'd started earlier, it would have made such a huge difference now! Yet you say that economists like William Nordhaus used discounting to justify not investing as much in fighting climate change ...

Hope: That troubles me too – especially that Nordhaus ended up winning a Nobel prize in economics for his climate work.

Caitlyn: (*shocked*) Really? That makes me sick to my stomach.

Hope: I know, honey. (*Hope stands up and gives Caitlyn a squeeze*) You know, between this heat and all this talk of ice cream, my sweet tooth is calling. Care to go inside for a slushie? It might help your stomach.

Caitlyn: Okay. (*smiling slightly*)

Part 2: Climate Change as a Challenge to Our Ethical Concepts

Ten minutes later, Hope and Caitlyn sit on high-chairs at the granite kitchen top, enjoying their slushies.

Harm and the Non-Identity Problem

Hope: We've talked a lot about problems in the way conventional economics deals with the future. But it's only fair to be clear that there's trouble in the philosophy too, particularly within ethics.

Caitlyn: Really? I would have thought the basic ethics was pretty easy. Don't inflict severe harm on people in the future for the sake of luxuries for yourselves. Don't threaten them with famine, mass starvation, social collapse, extinction. That sort of thing.

Hope: Well, you have a point.

Caitlyn: (*smiling*) Thought so.

Hope: How about we start here: Do you want to hear a puzzle involving intergenerational ethics that philosophers have thought hard about? I've always found it fascinating, although I warn you the challenge might also be a bit frustrating. It involves the concept of harm.

Caitlyn: I'll be fine. Shoot.

Hope: Let's start with a question. Do you believe that it is possible that a policy of, say, geoengineering the climate through stratospheric sulfate injection[31] could *harm* people living far out in the future, say 200 years or more from now? (*Caitlyn nods, still busy enjoying her treat*) Well, philosophers have posed a challenge to that belief.[32]

---

31 See Dialogue 5.
32 Schwartz, 1978; Kavka, 1981; Parfit, 1983a (see annotated bibliography); for critical discussion, see Woodward, 1986.

Caitlyn takes her attention away from the slushie long enough to portray a look of puzzlement.

Caitlyn: What do you mean?

Hope: Bear with me here. Imagine that Samantha, a young woman of eighteen, decides she would like to have a baby. Her mother grows concerned and tells Samantha that it would be worse for her future if she had a child now, rather than waiting until she was out of school and more financially stable. Samantha understands her mother's point but reminds her that even if she will be worse off, that's ultimately her choice.

Caitlyn: (*resolutely*) She's right.

Hope: So her mom realizes. Therefore, she tries a different tack. She argues that it would be worse for the *child* if Samantha were to have them now rather than later.

Caitlyn: I suppose there's something to that.

Hope: Well, even so, Samantha rejects her mother's advice. She has a child, Henry, and as it turns out Henry indeed has a poor start in life.

Caitlyn: This story is sad.

Hope: (*softly*) Nevertheless, the story raises a philosophical question I'd like you to think about: Was her mother right that Henry was worse off because Samantha didn't wait? Think about it for a few moments.

Caitlyn: I don't think Samantha did anything bad, really. But I suppose her son would have been better off had he been born after she was more financially secure.

Hope: Perhaps. But would *Henry* have been better off if Samantha waited?

Caitlyn: What do you mean?

Hope: Well, had Samantha waited, *Henry* would not have been born. Instead, she would have had an entirely different child.

Caitlyn: Oh. I see!

Hope: So, with this in mind, do you think that *Henry* is worse off because his mother had him at a young age?

Caitlyn: Interesting. I suppose not. I mean, Henry would not have existed at all if Samantha would not have had him when she did. Right?

Hope: Exactly.

Caitlyn: Wait! I suppose Henry could be worse off existing than not existing. I mean, if his life is so bad as to be *not worth living*. If he were in constant, excruciating pain, for example, that made his life a torment to him.

Hope: How awful. But let's assume that's not the case. This story isn't as sad as that. Henry's life is difficult in some ways, but not so bad as to be not worth living. So, for example, he values his life, and the hardships that come with it don't undermine that.

Caitlyn: Okay.

Hope: So, given this situation, and especially that Henry wouldn't exist if Samantha had chosen to wait to have a child, do you think her choice to have him *harms* Henry?

Caitlyn: No. I suppose Samantha didn't really harm him. As you say, he wouldn't exist otherwise, and we're assuming his life is worthwhile for him. Actually, that makes sense to me now I ponder it some more – I certainly don't think it's wrong to have a child at a young age if that's your decision.

Hope: So far, so good. But let me pose my original question again: Do you think that the United States could adopt a policy that would make people 200 years from now worse off?

Caitlyn: (*surprised*) Well, yes. Why wouldn't I?

Hope: The thought is that the same problem applies. Think about it this way. Suppose we were lazy and delayed decarbonization for another few decades. But then we deployed geoengineering – SSI in this case – to hold off the global temperature rise over the next century or so.

A policy that influential would dramatically affect who exists 200 years from now. By changing the world so much, it would alter which couples get together, when they conceive their children, and how they are raised. As this effect snowballs, over time – say 200 years in my example – it will turn out that everybody living in the geoengineered world is a different person from anyone who would have lived in a non-geoengineered world, where we were not lazy and didn't delay decarbonization.

Caitlyn: This sounds like sci-fi.

Hope: Maybe not. Something like it has probably happened in the past. Imagine how many of us would not exist if World War II hadn't happened. My grandparents – your great, great grandparents – wouldn't have met. So, presumably we wouldn't exist. I'm guessing that's true for a lot of people.

Caitlyn: Huh.

Hope: Anyway, do you see the philosophical problem? Suppose the geoengineering we initiate has nasty side effects that kill a billion people 200 years from now. It seems natural to say that these people were harmed by us, because they were made worse off by our choice to geoengineer. And yet, if we hadn't geoengineered, they wouldn't have *lived at all*. So, if they are like Henry – if their lives are still worth living – then, given what we said before, it doesn't seem like we actually harm them.

Caitlyn: Wow! That's really paradoxical. We seem to be saying that by adopting a genuinely dangerous geoengineering policy – one that eventually results in the avoidable deaths of a billion people in the future – we actually don't *harm* those people, since they would not have existed otherwise.

Hope: Indeed, some might even say we benefit them. For their lives are still worth living.

Caitlyn: Something's not right. Surely the dangerous geoengineering policy is still morally wrong if it results in that outcome.

Hope: Some think it is no longer morally wrong.[33] Still, most agree with you – including me. The challenge is explaining why. How could the choice to geoengineer be wrong if we know it will make nobody worse off?

Caitlyn: For one thing, couldn't our geoengineered world make people living right now worse off?

Hope: That could be true. Yet, in that case, the policy would be wrong because of how it affected already existing people. What we're trying to explain in our example is why we feel that we are doing something bad to the billion people in the future.

Caitlyn: I see.

Hope: To make this clearer, let's imagine a more general case. Consider two social policies we might adopt: one of depletion and one of conservation.[34] The policy of depletion will make everyone living within the next 200 years have a slightly higher quality of life compared to the policy of conservation. Afterwards, however, the quality of life becomes *significantly* lower, though still worth living. Which approach do you think current people should adopt?

Caitlyn: Conservation, *obviously*!

Hope: I agree. But notice that the depletion policy only reduces the quality of life for those living beyond 200 years from now, and they wouldn't exist at all *unless* we had adopted the very policy that seems to make them worse off. So, it seems like they aren't *really* made worse off, and therefore are not harmed. You might even say that they *benefit* from being brought into existence by the policy.

---

33 Cf. Boonin, 2014; Boonin, 2021.
34 Parfit, 1983a.

Caitlyn: (*groaning*) I see the point. Still, I'm not sure it makes sense to claim that future people have an interest in *existence* itself. [35] Existence itself is not a benefit – it is a precondition for being able to enjoy benefits, or suffer harms, at all. So, causing needless pain and death to future people is not justified by the "benefit" of bringing those billion people into existence because existence is not itself a benefit.

Hope: A fascinating suggestion. You may be onto something. Even so, we might still say that these future people in question – if they valued their lives at least – could not rationally condemn the depletion policy because they would not be alive without that very policy. Right?

Caitlyn: That doesn't sound right to me. Surely there are many ways in which you might disrespect people or otherwise wrong them. I'm not convinced that becomes irrelevant so long as you are causing them to exist at the same time.[36]

Hope: Interesting. Can you say more?

Caitlyn: I suppose a large part of my revulsion at depletion is that it produces a future world with people who are worse off than they ought to be.[37] It's wrong to cause people to live unacceptably diminished lives. I consider that disrespectful. I believe it robs them of their dignity. I'd call that a serious kind of harm. It is certainly a *wrong*, even if they wouldn't have existed any other way.

Hope: (*thoughtfully*) I like that answer. It's making me realize that there are different conceptions of harm. It seems like you have in mind a threshold conception of harm, where we ought not to cause a person to be worse off than that person is *entitled* to be. That's different from the conception of harm I was using, which involved the

---

35 Weinberg, 2008, pp. 13–17.
36 Kumar, 2003.
37 Meyer & Roser, 2009, p. 229. See annotated bibliography.

notion of the person being worse off than they *would be otherwise*. If I may ask, though, what threshold do you have in mind?

Caitlyn: For one thing, currently living people should not cause future people to live considerably worse lives than we do.

Hope: That makes some sense. But don't you think it's a little odd to assign value to *equality* in well-being, rather than actual well-being?

Caitlyn: Why?

Hope: If *relative* well-being is what truly mattered, then we might be justified in creating future people with very poor lives, as long as we ourselves were also living very poor lives.[38]

Caitlyn: Hmm. I do want to rule that out. I suppose I believe that everyone has a right to the opportunity to live a decent life. I guess what truly matters is that current people leave enough resources for future people to meet their needs, have a good quality existence, and exercise reasonable levels of autonomy and self-respect. The current generation has an obligation to ensure future people do not live below this threshold.[39] That applies whoever they are – I don't think their identity is particularly important.[40]

Hope: That's helpful, Caitlyn. I think your view does avoid the thrust of the non-identity problem, at least in many of the most important cases.

Caitlyn: Good. Anyway, I worry that this weird problem is just a cop-out. Muddying the waters so as to help your generation feel less responsible for the awful mess you created.

---

38 This is a form of the leveling-down objection to egalitarian conceptions of justice. Meyer & Roser, 2009, p. 220.

39 In the literature, this is known as a sufficientarian conception of intergenerational justice (e.g., Meyer & Roser, 2009, p. 220).

40 John Nolt argues that part of the wrong of a dangerous climate policy may rest in how it *dominates* posterity, and also that this sort of harm does not turn on the particular identities of the future people being dominated. See Nolt, 2011a.

Hope: So, you think my generation violated our intergenerational obligations to your generation?

Caitlyn: (*regretfully*) I do. You know better than me how climate change causes droughts, flooding, and extreme weather. People worldwide are starving from famine caused by droughts; they are displaced and forced to wander as refugees due to flooding; they die from extreme weather. Many in my generation are far below the threshold of decent well-being, and people your age and in prior generations caused this. It doesn't really matter if I wouldn't exist in a non-warming world – say, because you and grandpa would never have met without climate change.

Her voice broke. Hope scoots around the table and puts her arm around Caitlyn.

Hope: Are you okay, sunflower?

Caitlyn: (*wiping her eyes*) I don't know. Sometimes I just feel so powerless. It feels like my generation has been hurt by earlier people who should have known better. We have no way of escaping, and no way of holding them accountable for how they are harming us. After all this discussion, I still don't understand why more robust climate action wasn't taken.

The Concept of Responsibility

Hope: (*quietly*) My generation and the generations immediately before mine acted wrongly. While ours were not the first generations to hurt the future, we *were* the first to do so with such foresight and to such a terrible degree. We held incredible power over future people, but we abused that power. That being said, our actions were not malicious. We did not *intend* to cause harm.

Caitlyn: How much does it really matter if it wasn't malicious? If it was motivated by indifference, complacency, or wantonness, rather than fear, hatred, or greed?[41] The generations before mine knew enough. They were reckless towards the well-being of my generation, to the point of being callous.[42] Intention hardly matters: the wrong remains.

Hope: I certainly don't mean to deny the wrongdoing. We should have acted very differently. But I can't help but feel like we didn't fully understand what we were doing.

Caitlyn: What do you mean? Politicians and the public have known about the threat since the 1960s, right? At the very latest, from the 1990s. Back then there was still ample time to prevent this from happening.

Hope: I don't mean just scientific knowledge. Climate change was not like any other problem humanity had faced so far. Some called it "a perfect moral storm." As we talked about, there were many barriers to action: global and intergenerational, but also scientific and political and psychological.

Caitlyn: Sure – but it was clear that it was an ethical challenge, and your generations should have risen to that challenge.

Hope: Maybe we just weren't properly equipped to do so. Maybe the moral concepts handed down to us were not fit for purpose. They evolved in small communities and were directed towards immediate and tangible harms. They weren't suited for cases of many people doing small things over a long period of time that together amounted to unimaginable harm.[43]

Caitlyn: What do you mean?

Hope: Perhaps part of our problem was that our concept of responsibility failed us. Climate change was a radically new problem and humanity needed to revise our understanding of responsibility to solve it. You might say that,

41 Gardiner, 2012; Bendik-Keymer, 2012, pp. 266–269.
42 Gardiner, 2013.
43 Jamieson, 1992.

| | under the existing concept of responsibility, we were facing a situation where the global environment might be destroyed, yet no one would be responsible.[44] |

Caitlyn: I get the idea. (*hesitantly*) But Grams, don't you think it is a little *too convenient?* You said before that the greatest failure was moral. Now you're saying you weren't responsible?

Hope: No. I take responsibility, and so should my whole generation. Even I didn't do all I could. I consumed more than I needed. Perhaps my political strategy was too complacent. Maybe I had too much faith in existing political systems.

Caitlyn: So, why this stuff about the failures of the concept of responsibility?

Hope: It's an explanation, not an excuse. I don't think the climate crisis was caused by hatred or even laziness. Still, the problem demanded moral creativity and bravery like no other problem before it. We *should* have met the high standard asked of us, we *needed* to, but we didn't. We deserve blame, but I also think the young need to recognize how enormous the problem was that my generation was asked to solve.

Caitlyn: Hmm. I don't pretend to understand fully the challenge your generation faced. But the basic solution was known, even if it took sacrifice: your generation needed to wean yourselves off fossil fuels quickly, and you *could* have done that with enough effort. (*pauses*) I don't blame *you personally,* Grandma. I know you fought hard to mitigate climate change harms, and you did help the future with your fight. But others did not. And they *can* be blamed. My generation can never forgive your generation for its failure. I don't think later generations will either.

44   Jamieson, 1992, p. 149.

Hope: (*weakly*) All I'm suggesting, is that human morality was not equipped to meet the problem of climate change, and so we needed to revise our moral concepts.[45] That is a formidable task, one we have yet to complete. Even as posterity can and should hold us responsible for our failures, I hope you and future people might understand the great task required of us. And we have made some progress. For example, the CJC pushes forward the concern for marginalized communities in ways we direly needed earlier.

Caitlyn: I do see that as progress.

Hope: Moreover, among my fellow politicians, while it is true that I saw short-sightedness, narrow-mindedness, and cynicism, I also saw goodness. Alas, that goodness was not enough to face down a problem so great, at least all at once. (*sniffing and reaching for a tissue*) That is taking time. Too much time.

Caitlyn: It's okay, Grandma. I know this conversation is hard.

Hope: Thank you; but I'm fine.

Caitlyn: May I ask a question, then?

Hope: Of course.

Caitlyn: I don't understand why you think climate change challenged our moral concepts. Can you explain?

Hope: Sure. When I was a young woman, I remember a psychology professor saying that part of the issue was that there were no established moral rules against meddling in atmospheric chemistry, like there were around cruelty to pets, patriotism or sex.[46] Climate change rarely made us feel nauseated or furious or disgraced. Therefore, few railed against it, like many railed against eating kittens, flag-burning or gay sex. So, to see climate change as a problem of moral responsibility, we needed to revise our

45   Jamieson, 2010. See annotated bibliography.
46   Gilbert, 2006.

normal understanding of moral responsibility. And we didn't.

Caitlyn: Sorry, Grams. I don't follow you. Why would people not get riled up about climate change?

Hope: Well, think about it this way: certain things are paradigms of moral problems. These are the problems that humans have faced for as long as they have been human.

Caitlyn: Like what?

Hope: Here's an example. Suppose a prehistoric human takes the deer killed by another person out of spite, causing the victim's family to suffer from malnourishment. In this case, the person intentionally harmed another; the villain and the victim are identifiable, and they are both very closely related in time and space.[47] Socially, this behavior quickly becomes unacceptable. It receives blame and condemnation. And this translates to the modern-day. Consider Jack intentionally stealing Jill's bicycle. Again, Jack causes harm intentionally, the villain and victim are identifiable, and they are closely related in time and space. So, people quickly recognize, and *feel*, that what Jack does is wrong.

Caitlyn: I suppose I see what you mean. Some types of wrong action are common, and so perhaps are more obviously wrong.

Hope: But sometimes wrong actions aren't like these cases. Consider Jack and Jill again, but where none of the paradigm features hold. What if, acting independently, Jack and a large number of unacquainted people set in motion a chain of events that causes a large number of future people who will live in another part of the world from ever having bicycles. Do you still think what Jack does is wrong?

Caitlyn: Well, yes!

47   Jamieson, 2010, p. 436.

Hope: You may be right. But do you think Jack is acting *as badly* as he does when he intentionally steals Jill's bike?

Caitlyn: Well, no; I suppose.

Hope: I think the reason you feel that way is because it's such an unusual case. But so is climate change. Climate change happened, and continues to happen, partly because people set in motion forces that harm future people. Even if we can understand that climate change is bad, we do not feel the same moral urgency to stop it, because it is so different in structure to more familiar, paradigm moral problems, like Jack stealing Jill's bike.

Caitlyn: Climate change is an unusual problem. I get that. But I think our moral concepts are well equipped to grasp its severity.

Hope: You weren't convinced by my Jack and Jill examples, then?[48]

Caitlyn: (*a pause*) Not really. I don't think they're close enough to climate change. For instance, climate harms are far worse than losing your bike or never having access to bikes; and there's nothing in the second example about Jack's motivation for participating in that chain of events.

Hope: Can you think of a better analogy, then?

Caitlyn: How about this? Let's start with a different case.[49] Suppose George steals Sanjay's smoke alarm; then he sets fire to Sanjay's house while Sanjay is asleep inside; George does this because he is bored and would like a little excitement. I think that's at least as much of a paradigm case of behavior that is deeply morally wrong as Jack and Jill. In fact, I think it is more so. It is definitely more morally serious.

---

48 The following discussion draws from Gardiner, 2011b (see annotated bibliography). For further development of the debate, see Jamieson, 2013a and Gardiner, 2017a.

49 Gardiner, 2011b, pp. 43–44.

Hope: Wow! I agree that your example feels much more morally serious (*smiling weakly*).

Caitlyn: I also think it is closer to climate change. For instance, the harm being threatened – being burned alive – is more serious than losing your bike. So, that's like climate change. For another thing, the George and Sanjay example gestures toward the idea that many people decide to burn fossil fuels not out of need, but for the sake of luxuries, convenience, or boredom.

Hope: Okay. But that example is parallel to my first Jack and Jill case, where Jack steals Jill's bike. The individual intentionally harmed another person; the villain and the victim are identifiable; and they are both very closely related in time and space. Climate change isn't like that.

Caitlyn: Agreed. But starting with a stronger paradigm case makes a difference. Suppose we follow the pattern of the Jack and Jill examples by changing the George and Sanjay case like this. Acting independently, George and a large number of unacquainted people set in motion a chain of events that causes a large number of future people who live in another part of the world never to have smoke alarms, and to have their houses set on fire.

Hope: Hmm. I see your point. The new example seems less bad than the original George and Sanjay case. But I agree that still sounds pretty bad; and much worse than Jack and Jill's parallel case.

Caitlyn: Right. My wider point is that I think everyday morality going back a long time could understand that what George was doing is wrong.

Hope: If climate change was like your second George and Sanjay example, I agree a revision of everyday morality would not be needed.

Caitlyn: But ...?

Hope: I don't think climate change is like that example. For instance, there was – and continues to be – scientific uncertainty surrounding how and when climate change will harm future people. So, unlike your example, our behavior was *risky* but does not straightforwardly cause harm like your example.

Caitlyn: (*hesitantly*) That is a difference.

Hope: Another important disanalogy is that, as I said before, my generation did not *intend* to harm yours and future generations. We did not take pleasure in knowing that our actions would cumulatively starve children, tear communities from their historical homeland, or make weather more extreme and lethal. Many of us knew these were possibilities, but we did not want them to come about.

Caitlyn: Okay. But, to be honest, I'm not sure whether those disanalogies are big enough to make the ethics much less clear. However, suppose I concede for the sake of argument that they *might*. Nevertheless, some other disanalogies actually make my initial example seem more morally severe than climate change. Accounting for them may make climate change even more clearly immoral.

Hope: What do you have in mind?

Caitlyn: Climate change doesn't only harm future people or those in other parts of the world. Many in your generation are still alive, and very well may be reaping the consequences of failing to mitigate and adapt. And, of course, plenty of people in the United States – indeed, right here in Washington – are suffering from this sweltering heat with no place to seek shelter. Some die from it.

Hope: You're right. By now, climate change impacts everyone, and there are people all over the world who are especially vulnerable.

Caitlyn: I also think that climate change can't be said to have been caused by a series of isolated incidents like my first George and Sanjay example and your first Jack and Jill example suggested. Climate change is happening because of long-term *patterns* of action. It has been caused by many individuals, who are also part of countries who themselves have contributed to the problem through their policies. Moreover, these countries are a part of an international community that has publicly accepted the reality of climate change since the early 1990s and repeatedly promised – but failed – to address it.

Hope: And you think all that makes the behavior worse?

Caitlyn: I do. It suggests that many people and institutions are implicated. They are complicit, or at least guilty by association. And in a way that the first examples didn't capture.

Hope: (*shudders*) I see. Well, is there an example that accounts for these things?

Caitlyn: Probably no analogy is perfect. But let's try pushing the George and Sanjay case a little further. Suppose George and his buddies frequently have big firework displays over poor neighborhoods, which increasingly imposes a serious risk of the houses in those neighborhoods catching on fire. They are aware of this risk and keep saying they will cut back, but never do. In fact, they keep making their fireworks bigger. (*Caitlyn looks at Hope seriously*) I think *that* is like what your generation did to the future, and anybody should be able to see that it was very morally wrong.

Hope: (*sighing*) I have to say I'm still not fully convinced of that analogy. For one, there is no intergenerational aspect to that example, and I think that was one of the largest obstacles to action.

Caitlyn: But shouldn't governments represent the interests of their future citizens? If so, that moral difference should not matter. And, if governments are not representing the interests of the future, current citizens can be held morally responsible if they do not create institutions that *do*.

Hope: I hear what you're saying, sunflower. I agree with much of it. My point is only that to have succeeded we needed to revise our concepts of individual and collective moral responsibility from what they've been historically.

Caitlyn: (*in frustration*) I'm sorry, Grams. I can't accept that. As far as I can tell, there's not much wrong with the concepts; there's a lot more wrong with the people. Past generations, including yours, should have known they were responsible. I suspect they did know; you did, after all.

Hope is quiet, obviously distressed.

Caitlyn: Grams, I know you always try to see the best in people, and to excuse them their failings. But I really think this is too big. You know that I love and honor you personally. You tried to do something – indeed, you dedicated much of your life to it – but most did not. In fact, your generation as a whole – and the one before – let us down. Ultimately, this disaster unfolded on their watch. They knew they were failing us; but they did it anyway, no matter how many clever and self-indulgent lies they told themselves to excuse it. Right now, I can't forgive them. Perhaps I should. Maybe I will be able to eventually. But it's asking a lot. I'm not ready and they shouldn't expect it of me. I'm sorry if that sounds extreme; maybe it is even a little unfair. But I can't see my way past it right now.

Hope: (*tearing up*) Sadly, I understand. I fear that we deserve your censure.[50] In a way, I always have. It is one of the things that pushed me forward. I worried about what future generations would think of us, my generation and the one before. I felt anguish that they would look back and see us as tarnished – in fact, worse than that, *blighted*. I mean that they would see the wrongs we've committed as irredeemable, as things that could not be

---

50 Following Gardiner, 2010, p. 301; Gardiner, 2012.

outweighed by other good things we might have done. Evils in that sense. Other generations have received praise from their successors – like the so-called greatest generation that tackled the Great Depression, World War II, and the rebuilding afterwards. My fear has been that we will fare much worse, and perhaps deservedly so. Perhaps future generations will even think of us as "the Scum of the Earth."[51]

Caitlyn: (*softly*) Oh, Grams. I'm so sorry. ... For whatever it is worth, I wouldn't go *that* far. And I certainly wouldn't put it on you, personally. You're one of my heroes ...

Hope and Caitlyn sit in silence for a few minutes, lost in their own thoughts and emotions.

Respect for Nature

Caitlyn: I do believe that humans probably need to rethink some of our moral commitments. If our moral and political systems have allowed dangerous climate change to develop, clearly something went very badly wrong.[52] Humanity needs to learn how to live well on the planet, and *in relationship to* the planet, even as we grow and develop.[53] It's a basic evolutionary challenge; and a basic test for our systems too. But we're failing; we've fouled the nest.[54] To meet humanity's challenge, what must change?

Hope: (*slowly, recovering herself*) That's a hard question. For a long time now I've believed that part of the reason the last few generations failed to respond adequately to climate change was our lack of respect for nature.[55] Even

51 Gardiner 2012.
52 Gardiner 2011a, chapter 7.
53 Gardiner, 2011b, p. 56.
54 Gardiner, 2012.
55 Jamieson, 2010, p. 440.

before I won my seat in the House, almost one half of Earth's land surface had been transformed by human action, we'd used half of the world's fresh water, and we'd driven a quarter of bird species to extinction.[56] (*Hope shakes her head*) This is one reason why I get so excited seeing the amazing, beautiful work of the CJC in protecting nature.

The older woman walks over to the sink and starts scrubbing dishes absent-mindedly.

Caitlyn: I understand what you mean. (*she grabs a towel to help with drying*) Still, I've heard people say humanity got as far as we did by *mastering* nature. Electricity, air travel, (*she runs a finger through the faucet stream*) running water. Genuine human freedom and flourishing, these people say, requires command over nature.[57] So, I wonder, is there really a duty to respect nature?

Hope: Humans dreamed of mastery long before I was born.[58] I think it's short-sighted. Experiencing wild nature helps give our lives meaning, grants perspective, and sustains psychological wellness.[59] I sit on this porch as much as I can, watching the birds and listening to the whisper of the wind through the trees to remind myself that nature is still there. I do this to remind myself that it's not too late to realize a dream of *naturalism*, where we do not attempt to force nature to be how we want it to be.[60] We preserve nature's wildness, and allow nature to nourish and transform us. Instead, so many people want to address our problems with big technological interventions: more domination.

---

56 Vitousek, *et al.*, 1997; Jamieson, 2010, p. 440.
57 Wapner, 2010, p. 81.
58 Wapner, 2010, chapter 4.
59 Jamieson, 2010, pp. 442-443; Frumkin, *et al.*, 2017.
60 Wapner, 2010, chapter 3.

Caitlyn: I agree that humanity is short-sighted. But maybe the problem is that we've been stuck with two starkly opposing visions: naturalism *or* mastery. Isn't there a third path?[61]

Hope: What do you have in mind?

Caitlyn: I'm wondering about some kind of enlightened ethical management.[62] Why not acknowledge that the future of life on Earth *depends* on us, but aim at managing nature with benevolence, generosity, and wisdom rather than domination? We've seen what good species relocation and ecosystem design can do.

Hope: We've also seen such interventions go awry. Humans are so often arrogant. In our technological arrogance, we think we know enough to successfully dominate nature; in our moral arrogance, we think we have the *right* to reign.[63] I think both attitudes are terribly wrong. We know far less than we think, and nature should not always bend to our will.

Caitlyn: Your vision of harmony *with* nature is beautiful, Grandma, but I can't help but wonder if its time has passed. We both agree with the CJC's commitment to care for nonhuman animals and Earth's ecosystems. But, to do right by nonhuman creatures and their communities, I believe we must intercede. After all the harm we've done, don't we owe them at least that?

Hope: Your heart is in the right place, love. But even the best intentions can result in ecological chaos, and humanity's intentions are very often not so good. I've seen it. Still, you make me *want* to believe in humanity's ability to build an ecologically just and flourishing world.

---

61   Wapner, 2010, pp. 205–219.

62   This is our term. J. Baird Callicott (1991, p. 361), Allen Thompson (2009, p. 97), and Emma Marris (2011, p. 171) all endorse something like enlightened ethical management, so described.

63   Obst, in preparation 2.

Hope and Courage in a Warming World

Caitlyn: I believe with all my heart that it is possible. The CJC has done such wonderful work pushing our governments to accept responsibility for the ecological crisis we have created. I believe this will continue.

Hope: (*smiles tearily*) To think otherwise would be to think terribly little of ourselves, wouldn't it? These past fifty years have been full of failure, but there have been some successes. I have faith that your generation will navigate us to many, many more – that the next hundred years or so will be an era of repair, and healing.

Caitlyn: Is that what helped you keep fighting this whole time? Is that how you've remained hopeful?

Hope: In part, but I also saw no other choice. Some say it's a moral responsibility to sustain hope. After all, justice is never given freely; so, pursuing justice always requires a willingness to seek justice even in circumstances that are antagonistic to its creation.[64] If I despair and believe justice to be impossible, then perhaps I have no reason to demand it. But we *must* demand justice.

Caitlyn: I've never thought about hope like that before.

Hope: In my life, I've watched as many gave up hope too quickly. For a long time, people refused to believe that climate change was happening at all. Then, when climate change became undeniable, they swiftly descended into despair.

Caitlyn: People denied climate change was occurring because in their hearts they knew that they couldn't justifiably continue to act the way they wanted. So, when they could no longer deny, they despaired. To me, the slide between denial and despair among the privileged and powerful is way too easy. It smacks of moral corruption.[65]

---

64 Moellendorf, 2006.
65 Dout & Obst, forthcoming.

Hope: I don't think we should be too harsh, sunflower, espe-
           cially on everyone. Some may be morally corrupt; but
           others are simply sensitive, even fragile. We must be
           kind to them, support them. They are still our sisters
           and brothers after all.

Hope nodded, staring off into rich hues of the horizon.

Caitlyn: It wasn't the sensitive I was complaining about. If any-
           thing, we need more sensitivity in the world, not less.

Hope: Amen to that. Still, I think you are right that we also
           need to cultivate resilience. We can be sensitive to the
           trouble, and to the loss. But we must also move forward;
           we must not let loss cause us to give up. That's partly
           why my own view is that courage is the most important
           virtue here, not hope. As I see it, we have a duty to be
           brave and face the world we are in.

Caitlyn: But what if courage is not enough to get us to the change
           we need?

Hope: There are no guarantees. But it was ever this way. Few
           of the challenges of life come with them. Nevertheless,
           if you want my thoughts, I'd say that even if we see very
           little possibility that it will all work out – indeed, even if
           we *cannot* see a way through – then we must still do our
           very best to be present, to bear witness. We must fight
           for the future, even if it seems futile. We owe it to the
           future, to each other, and to ourselves.

Caitlyn: You may be right that courage is just as important as
           hope. Yet, to me, they feel intimately intertwined. The
           other day in school I learned about refugees all around
           the world who've had to leave their homes due, in part,
           to climate pressures. I tried to imagine myself in their
           shoes. How I would feel if I lost my home? Could I even
           hope for a new home, having no idea where it would be,
           or what my life would be like in a new country? How
           can a person find a new, radical hope that they will have

a good life in an unknowable future?[66] I think it must take tremendous courage, don't you?

Hope: (*softly*) I do. Maybe that's the kind of radical virtue we need now. Our warmer world is in flux, and lives will change. In truth, I can hardly imagine what human flourishing might look like when you're my age, child.

Caitlyn: I know warming will continue, and I know there will continue to be many obstacles to justice. Still, I also know there will be those who will stand up for what's right, as you and others have done. As I intend to do.

Hope looks sad but says nothing.

Caitlyn: I believe that my generation and those to come will need to embrace radical hope. We *must* find ways to live good lives in this new, unstable climate as it unfolds before us.[67] We *must* seek out that future with courage.

Hope: (*wiping a tear from her eye*) So, perhaps you need courage even to exercise radical hope?

Caitlyn: (*smiling*) Hope or courage; the chicken or the egg, huh?

66 Lear, 2008; insightfully applied to climate change in Thompson 2010 and Shockley 2020. Lear discusses radical hope through the real-life example of Chief Plenty Coup, who in the nineteenth century led the Crow through a time of great cultural dislocation, caused by severe injustices committed against them, particularly by the U.S. government. The Crow were forcibly moved away from their homelands, and this disrupted much of their traditional way of life, including by making buffalo hunting impossible. Seizing on Plenty Coup's remark that "after that nothing happened", Lear explores the possibility that the Crow experienced *total cultural annihilation*.

This example is now common in the literature. However, we have chosen not to emphasize it here. While the severe losses and great injustices suffered by the Crow are not in doubt, we worry about the accuracy of the more specific suggestion of total cultural annihilation, and the distress it may cause the Crow and other indigenous peoples. We also have concerns about the comparisons sometimes made between Plenty Coup's situation and some of the more moderate sacrifices facing more affluent communities in dealing with climate change (e.g., the loss of high consumption lifestyles). We have therefore decided to use a different, less-specific example with broader application.

67 Thompson, 2010; Shockley, 2020.

Hope: (*returning her granddaughter's smile*) Maybe we need
           both virtues, feeding off one another. In graduate school,
           I remember reading Rebecca Solnit, a writer and activ-
           ist. Some of her words have stuck with me ever since:
           "Hope is an ax you break down doors with in an emer-
           gency; because Hope should shove you out the door,
           because it will take everything you have to steer the
           future away from endless war, from the annihilation of
           Earth's treasures and the grinding down of the poor and
           marginal." (*Hope's words hang in the air between them
           for a long moment*)[68] I've always loved that quote. To
           me, it captures something about hope when it is work-
           ing well: its vivacity, its stubbornness, its courage. Hope
           isn't, or shouldn't be, some kind of passive virtue. It has
           real energy, even fight, to it.

Caitlyn: Well said!

Hope: I've tried to live my life this way. I pushed for change
           and I hoped, as bravely as I could. Often, no doubt, that
           was not bravely enough. I still fear that I've failed you …
           (*Hope collects herself and looks over at her granddaugh-
           ter*) Still, I cling to the dream that ultimately you'll find a
           pathway through. My darling, never give up that radical
           hope of yours. Never lose your courage.

Caitlyn stares back, eyes the color of the sunset.

Caitlyn: I won't.

---

68   Solnit, 2004.

Some Years Later

Caitlyn: Grams! Hurry up or you'll miss it!
Hope: (*A little way off*) Coming …!
Caitlyn: (*Excited*) I can see Mom … Auntie Eliza and Uncle Andrew are there too … and Grandpa Carlos! They all look so different in their fancy clothes!
Hope: (*Slightly breathless*) I'm here now.
Caitlyn: It's the big moment!
Hope: Yes. I feel like I've been waiting for most of my life for this day. … (*pauses, tears in her eyes*) … Hold my hand, sunflower. … Hold it tight ….

Appendix

Character List

(In Order of Appearance)

Hope: Law Student; U.S. Negotiator; U.S. Congressperson; Grandma.

Eliza: Graduate student, economics and policy.

Stacy: Graduate student, climate science.

Nadia: Stacy's best friend, designated non-expert.

Andrew: Activist, Climate Justice Now.

Adama: Activist, the Anti-Colonial Alliance.

Dad: Hope's dad.

Richard: Negotiator, U.S.

Susan: Climate negotiator, U.K.

Amina: Climate negotiator, Maldives.

Indira: Climate negotiator, India.

Marcia: Climate negotiator, Brazil.

Carlos: Climate negotiator, Bolivia.

Kim: Climate negotiator, China.

Peter: Climate negotiator, poverty NGO.

Joseph: Climate negotiator, Democratic Republic of the Congo.

Ernest: Geoengineering expert, physicist-turned-engineer.

Thomas: Geoengineering expert, expert on Earth systems.

Caitlyn: Hope's granddaughter.

Annotated Bibliography

Suggested readings in order of their appearance in the text.

Dialogue 1: Why Ethics?

Intergovernmental Panel on Climate Change. (2021). Summary for Policymakers. In *Climate Change 2021: The Physical Science Basis*. Cambridge University Press.

> In 2021, Working Group I of the Intergovernmental Panel on Climate Change released an updated report on the physical science underlying climate change. This assessment forms part of the Sixth Assessment Report (AR6), which builds upon the findings of earlier assessments in 1990, 1995, 2001, 2007, 2014, and special reports released in 2018 and 2019. The new report states that human-caused climate change is "unequivocal" (6), humans have warmed the climate at a rate "unprecedented in at least the last 2000 years" (7), and atmospheric CO_2 is higher than any time in "at least 2 million years" (9). In addition, there is updated evidence that climate extremes – including heatwaves, heavy participation, droughts, and tropical cyclones – are attributable to human influence (10), and have already affected every inhabited region across the globe (12). AR6 models five future emissions scenarios and concludes that 1.5°C warming is *more likely than not* to be exceeded on all but one (18). In summary, the report confirms what almost everyone working on climate already knew: that in the years since AR5, the climate crisis has

accelerated, and invaluable time has been lost in meeting the international targets agreed in Paris in 2015.

For discussion: Reflecting on this piece, you might ask yourself, what parts of the science make the biggest difference to how you think about climate change? How might they bear on the questions of ethics and justice this book considers?

Sheila Watt-Cloutier. (2010). The Inuit right to culture based on ice and snow. In K.D. Moore & M.P. Nelson (Eds.), *Moral Ground: Ethical Action for a Planet in Peril*. Edited by San Antonio: Trinity University Press.

Canadian Inuit activist Sheila Watt-Cloutier highlights the ongoing harms that climate change inflicts on her people and their culture, as it plays out against a backdrop of historical traumas. Some of the many impacts include that warmer weather melts the highways of ice and snow, erosion eats at the landscape, the coastline slips into the sea, foreign toxins permeate their food, and many native animals face extinction. Nevertheless, Watt-Cloutier urges the reader not to get too caught up in thinking about climate change as a scientific and technical problem. Instead, we must recognize it as a *human* issue that should be confronted through the lens of human rights. The very way of life of Inuit communities is under threat. We should, she says, respect her people's "right to be cold."

For discussion: In addressing this article, you might consider how Watt-Cloutier's arguments affect how you should think about climate impacts in places you care about, how you would develop this kind of view, and whether you have any concerns about it.

Garrett Hardin. (1968). The tragedy of the commons. *Science* 162(3859): 1243–1248.

Ecologist Garrett Hardin popularized the "tragedy of the commons" as an explanation for environmental depletion.

Elucidating an idea he draws from William Forster Lloyd (1833), Hardin argues that an unregulated pasture open to all herdsmen will result in overgrazing provided that – as rational actors – each herdsman seeks to maximize their individual gain. In short, in an open pasture, acting rationally in an instrumental sense results in collective collapse. Hardin then extends this analysis to pollution and overpopulation, suggesting that the perverse incentives present in the open pasture problem exist in these other familiar contexts. Hardin finally argues that appeals to conscience or morality will not solve these commons problems and presents an alternative solution: mutual coercion, mutually agreed upon.

For discussion: In reading this article, you might consider whether you think Hardin's "tragedy of the commons" framing is appropriate for understanding the roots of the climate crisis. Whether you think it is or not, how does your attitude to Hardin affect what solutions you'd like to call for?

Elinor Ostrom. (2009). A general framework for analyzing sustainability of Social–Ecological Systems. *Science* 235: 419–422.

The tragedy of the commons model of social–ecological systems typically assumes that resource users will deplete a commons rather than self-organize to ensure sustainability. To avoid ecological degradation and collapse, the dominant story goes, we should demand government regulation of fisheries, lakes, forests, and the atmosphere. In this article, economist Elinor Ostrom challenges this approach. According to her extensive interdisciplinary research, complex social and ecological factors make prescribing successful management strategies more difficult and context-dependent than Hardin, for example, suggests. Social norms are one especially important, but oft-overlooked, variable: for instance, "users ... who share moral and ethical standards regarding how to behave in groups they form, and thus the norms of reciprocity ... will face lower transaction costs in reaching agreements and lower costs of monitoring" (421). Such normative factors, which Hardin dismisses, may undermine his view that external coercion is needed to solve commons problems.

For discussion: In contemplating Ostrom's account, you might consider: Is moral transformation a preferable strategy for preventing commons' collapse? What might such a transformation look like?

Stephen Gardiner. (2016). Betraying the future. In S. Gardiner and D. Weisbach, *Debating Climate Ethics*. Oxford: Oxford University Press.[1]

In this updated version of his view from *A Perfect Moral Storm: The Ethical Tragedy of Climate Change* (Oxford, 2011), philosopher Stephen Gardiner argues that ethics plays a fundamental role in climate policy for three reasons. First, we need ethical concepts to identify the relevant problem, otherwise we risk violating intelligibility constraints. Second, ethical considerations are at the heart of the main policy decisions that must be made, such as how quickly to reduce greenhouse gas emissions over time, how those emissions that are allowable at a given time should be distributed, and what should be done to address unavoided impacts. Third, and more generally, climate change poses a severe ethical test to humanity and its institutions. Specifically, Gardiner argues that climate change is a perfect moral storm, which brings together serious challenges to ethical action at the global, intergenerational, ecological, and theoretical levels, and also encourages moral corruption. Gardiner claims that until we acknowledge and face the perfect moral storm, our solutions are likely to prove shallow and hollow.

---

1 As alternatives, instructors might also consider the original article version of the perfect moral storm (Gardiner, 2006), or the more detailed version of Gardiner, 2011, chapters 1–2. We recommend 'Betraying the Future' in part because it explicitly considers the ecological storm.

For discussion: In reflecting on this view, you might ask your-self what parts of the perfect moral storm seem most important to you, and whether they are obscured by other analyses. You may also consider whether moral corruption is a real problem, and whether you see examples of it in contemporary discussion of climate change.

Clare Heyward. (2014). Climate change as cultural injustice. In T. Brooks (Ed.), *New Waves in Global Justice*. London: Palgrave Macmillan.

When the UNFCCC formed in the mid-90s, climate mitigation – the reduction in global GHG emissions to reduce global warming – was the center of concern. However, since then, as mitigation efforts have stalled, climate adaptation has been increasingly central to climate change responses. Whereas mitigation seeks to keep the world cool, adaptation aims to conserve key human interests in a warmer world. In policy, these interests are often understood primarily in economic and material terms. In this chapter, philosopher Clare Heyward argues that human interests cannot be so narrowly conceived. In particular, she claims that climate change threatens cultural identity in morally significant ways, and that adaptation policy should be designed to protect against this loss and injustice. Finally, Heyward explains why the moral value of cultural identity cannot justify an "American way of life" dependent on high GHG emissions.

For discussion: Heyward's paper emphasizes the moral values embedded in the diversity of human experience, and invites a profound question: Which ways of life ought to change in a warmer world, which should be conserved, and how do you tell the difference?

Kyle Powys Whyte. (2019). Way beyond the lifeboat: An Indigenous allegory for climate justice. In K.K. Bhavnani, J. Foran,

P.A. Kurian, & D. Munshi (Eds.), *Climate Futures: Re-imagining Global Climate Justice*. London: Zed Books.

> Environmental ethics contains a complicated history of ship metaphors. In the 1960s, Buckminster Fuller imagined the planet as a spaceship without the ability to acquire more resources. In the 1970s, Garrett Hardin described the environmental predicament posed by rapid population growth as one where rich countries should be seen as analogous to lifeboats about to be swamped by poor people escaping overpopulated countries that were sinking. In this paper, philosopher Kyle Powys Whyte argues that such analyses fail to capture the complex context of environmental change, and in particular the relationship between colonialism, capitalism, industrialization, and environmental vulnerabilities. To make this more visible, Whyte offers a rival allegory (15) that draws out what we think of as the skewed vulnerabilities, historical responsibilities, and background injustices that previous ship metaphors missed. Whyte argues that this new allegory demonstrates how "in the absence of a concern for addressing colonialism, climate justice advocates do not really propose solutions to climate change that are that much better for Indigenous well-being than the proposed inaction of even the most strident climate change deniers" (20).
>
> *For discussion*: Whyte's conclusion encourages the idea that genuine climate justice must pay special attention to historical and political context, and to the voices of a diverse range of communities. In thinking through this idea, you might consider whether you have encountered climate policy proposals that seem vulnerable to Whyte's criticisms, and what might be done to improve them.

Dialogue 2: Skepticisms

Naomi Oreskes & Erik Conway. (2010). The denial of global warming. In *Merchants of Doubt: How a Handful of Scientists*

Obscured the Truth on Issues from Tobacco Smoke to Climate Change. London: Bloomsbury Press.

In a book that proved successful across both academic and public audiences, historians Naomi Oreskes and Erik Conway highlight various industries and individuals that have obscured the reality of public health threats, often for profit. In chapter six, they turn their attention to climate change. They begin by arguing that there was a broad scientific consensus as early as the mid-1970s that global warming would occur. Then, they document the ways in which a few influential actors that downplayed or denied this scientific consensus in the years to come. Ultimately, Oreskes and Conway suggest that these actors were likely motivated by their philosophical affinity for small government, and contend that they were aided by a mass media who cultivated the inaccurate image of controversy under the justification of journalistic "balance." Their book argues that public perception can be deeply influenced by hidden networks of power, and suggests that this phenomenon can have deeply damaging effects on public health and the environment.

For discussion: Reflecting on this chapter, you might ask yourself what you think the appropriate norms should be for public discussion of environmental issues, and whether current discussion of climate change respects such norms. If they don't, how ought they to change?

Catriona McKinnon. (2016). Should we tolerate climate change denial? *Midwest Studies in Philosophy* 40(1): 205–216.

Most believe that a central precept of a truly free society is a strong moral presumption in favor of tolerance: even if I do not like what you say, I should respect your right to say it. Therefore, they infer that there is a strong moral presumption in favor of tolerating even "the deliberate and deceptive misrepresentation of the scientific realties of climate change" (205). In this paper, philosopher Catriona McKinnon takes on this heavy burden of proof, and argues that a liberal society should be *intolerant* of this kind of climate denial. McKinnon makes her case by comparing climate denial to a famous limiting case to free speech: the act of shouting fire in a crowded theater. In such situations, she argues, silencing speech is justified because free expression stands in the way of the urgent and immediate action required to prevent extreme harm. By tweaking the theater example, she infers that the same can be said of limiting climate denial today.

For discussion: McKinnon's argument may be unsettling to some. Do you think it succeeds? Could this possible demand for illiberality in liberal societies be another tragedy of climate change?

René Descartes. (2008). Of the nature of the human mind; and that it is more easily known than the body. In *Meditations on First Philosophy: With Selections from the Objection and Replies*. Moriarty, M. (Ed./Trans.). Oxford: Oxford University Press.

In this classic of Western philosophy, philosopher René Descartes attempts to arrive at absolute and indisputable knowledge. His method is to purge himself of any beliefs that are subject to uncertainty by refusing to believe anything that he can possibly doubt. He quickly dismisses the reliability of his senses, noting that he often believes the reality of his senses in his dreams. He then considers those things about which he is more confident: for instance, the accuracy of simple arithmetic or the existence of a benevolent god. However, he then admits that it is possible that there exists an evil genius who is deceiving him about these things as well. Ultimately, he concludes that the only thing of which he can be completely certain is that he exists: after all, to doubt his own existence, there needs to be some prior thinking thing that is doing the doubting (in his famous phrase: "I think, therefore I am"). From this foundation, Descartes attempts to rebuild his confidence in other forms of knowledge.

For discussion: Descartes' philosophical strategy illustrates how almost *anything* can be doubted, and on a rational basis. But it also shows that the price of such radical skepticism is extreme: very little that we normally claim to know can survive. This suggests that adopting too skeptical an attitude to our beliefs is neither wise, pragmatic, nor ethical. In thinking about this point, you might ask yourself, what levels of skepticism are reasonable and desirable when it comes to scientific claims in general, and to climate science in particular?

Carl Hempel. (1960). Science and human values. In R.E. Spiller (Ed.), *Social Control in a Free Society*. Philadelphia: University of Pennsylvania Press.

It's tempting to think that some important social questions can be answered by the objective methods of science. After all, there *is* an empirical matter of fact behind the most effective way to curb population growth, whether tobacco smoke causes lung cancer, and the existence of anthropogenic climate change. In this classic work, however, the philosopher Carl Hempel argues that ethical values inevitably run through any scientific inquiry. He presses multiple arguments to this conclusion, but most importantly for our purposes he points out how accepting or rejecting a hypothesis always comes with "inductive risk": the chance of being wrong, and the consequences associated with it (92). In the climate context, one might say that the consequences of erroneously *rejecting* the climate change hypothesis are incredibly high – unmitigated extreme weather, sea-level rise, and loss & damage – and this must be considered when determining the evidential threshold for acceptance. On the other hand, erroneously *accepting* the climate change hypothesis might come with substantial costs from needless decarbonization. Importantly, how inductive risk affects the rules for accepting or rejecting a given hypothesis inevitably involves a non-epistemic, *moral* value judgment. Even the choice to *ignore* the consequences of inductive risk is morally laden. In this way, the choice to accept or reject the science of anthropogenic climate change cannot help but involve a moral judgment.

For discussion: Given arguments about inductive risk, reflect on what moral judgments you think are appropriate in the climate context.

Stephen M. Gardiner, & David A. Weisbach (2016). *Debating Climate Ethics*. Oxford: Oxford University Press.

In *Debating Climate Ethics,* Stephen Gardiner and David Weisbach offer contrasting views of the importance of ethics, self-interest, and justice for climate policy (Gardiner & Weisbach, 2016). Gardiner argues that "climate change is fundamentally an ethical issue" that "should be of serious concern to both moral philosophers and humanity at large," and that "the temptation to defer to experts in other disciplines should be resisted" (Gardiner, 2004, 556). By contrast, Weisbach claims that "ethics ... is not the right tool for the design of policies" and "trying to use philosophy to design climate change policy will, except by sheer happenstance, lead to bad policies" (151–2). Instead, he advocates climate policy based on simple, narrow forms of self-interest (149; 154): "policies based purely on self-interest – a desire to stop hitting ourselves in the head with a hammer" (197). Weisbach emphasizes the need to respect constraints of political feasibility based on national self-interest, and claims that mainstream proposals in climate ethics violate these constraints. Gardiner counters that climate policy driven by short-term, narrowly economic forms of self-interest threatens to undermine serious climate action, and may even encourage extortion of poor countries and future generations. For such reasons, he thinks ethics is necessary.

For discussion: Reflecting on this debate, ask yourself what roles you think ethics and self-interest should play in climate policy. Are ethics and national self-interest in conflict? What account of self-interest should we employ? Would it matter if climate action were successful, but also extortionate?[2]

Dialogue 3: Individual Responsibility

Walter Sinnott-Armstrong. (2005). It is not my fault: Global warming and individual moral obligations. In W. Sinnott-Armstrong, & R. Howarth (Eds.), *Perspectives on Climate Change: Science, Economics, Politics, Ethics* (pp. 221–253). Elsevier.

In the paper that spawned a vibrant literature on individual responsibility for climate change, philosopher Walter Sinnott-Armstrong considers whether one has a moral obligation not to emit reasonable amounts of greenhouse gases into the atmosphere just for fun, taking as his main example going for a Sunday afternoon drive in a powerful, gas-guzzling SUV just for the fun of it (subsequently, this became known as "joyguzzling"). Sinnott-Armstrong surveys several possible moral principles that might undergird a moral obligation not to joyguzzle and finds them all lacking. The largest reason for their failure, he contends, is that climate change will occur no matter how much one personally emits. After his writing, this becomes known as *the problem of inconsequentialism*. Sinnott-Armstrong anticipates his conclusion may upset many environmentalists but thinks that it illuminates an important lesson: individuals should not keep their hands clean at the expense of collective organization. In a memorable passage, he concludes: "Some environmentalists keep their hands clean by withdrawing into a simple life ... [but] they rarely come down out of the hills to work for political candidates who could and would change government policies. This attitude helps nobody ... It is better to enjoy your Sunday driving while working to change the law, so as to make it illegal

---

2 Instructors wanting to go into more depth might assign specific chapters from this book. Versions of some arguments also appear in Posner & Weisbach, 2013, and Gardiner, 2021b.

for you to enjoy your Sunday driving" (312). In some ways, Sinnott-Armstrong's point here was prescient: in the fifteen years since his writing, the debate between the importance of individual lifestyle change versus structural change has animated both academic and public discussions of climate ethics.

For discussion: What's your reaction to the memorable passage? Do you agree with Sinnott-Armstrong that it is not morally wrong to joyguzzle on a Sunday afternoon? What do you think is the strongest objection to your view? How would you respond to that objection?

Aaron Maltais. (2013). Radically non-ideal climate politics and the obligation to at least vote green. *Environmental Values* 22: 589–608.

In debates surrounding individual climate responsibilities, often the choice is framed between individual (unilateral) action and collective action. Here, philosopher Aaron Maltais disrupts this dualism by arguing that both individual and collective action suffer from the same general problem: one's personal actions will not make a difference. This is true of both individual emissions–reductions or contributions to collective action. Unsustainable and insular individual lifestyles are only immoral and unfair when others are likely to work cooperatively to ensure collective benefit: e.g., in the presence of large-scale collective efforts and when the global carbon budget is reasonably likely to be met. However, Maltais argues, sufficient others are *not* likely to cooperate in the climate case; we are in a radically non-ideal climate politic. As a result, one does not have a moral *obligation* to take on any individual or collective climate actions that entail significant personal burdens. Rather, one only has the obligation to "at least" vote green – an act made obligatory because of its tiny cost to the voter.

For discussion: Do you think you have an obligation to vote "green"? Is this the only individual obligation you have? What do you think the limits of acceptable personal burdens are?

Marion Hourdequin. (2010). Climate, collective action and individual ethical obligations. *Environmental Values* 19: 443–464.

In this paper, philosopher Marion Hourdequin responds to arguments against individual climate obligations. She grants that climate change typically has a tragic structure that undermines the efficacy of individual behavior change but maintains nonetheless that individuals have moral obligations to reduce their personal carbon emissions. She defends this thesis in two ways. First, she argues that someone who demands collective climate action but frivolously burns fossil fuels would be working at cross purposes, and therefore expressing the vice of hypocrisy. By contrast, the virtue of moral integrity demands that one harmonizes the commitments one holds in the various spheres they inhabit – whether personal or social. Second, drawing from Confucian philosophy, Hourdequin suggests that human beings define themselves relationally, and emphasizes that one's own choice to bike to work or go vegan might spur others to adopt greener behaviors. Thus, even without establishing any direct physical connection between an individual act of emitting and anthropogenic global warming, Hourdequin presents a serious argument for why the virtuous moral agent would live a low-emissions lifestyle.

For discussion: Do you agree with Hourdequin's concerns about integrity and hypocrisy? What do you think virtue and vice look like in these cases? Does this article make you think differently about your own behavior, or not? Why?

Christian Baatz. (2014). Climate change and individual duties to reduce GHG emissions. *Ethics, Policy & Environment* 17(1): 1–19.

When considering individual GHG reduction duties, philosopher Christian Baatz begins with a basic principle of equality: everyone has the right to emit the same amount of GHGs. To survive, every person must use the atmosphere as a sink to some extent, and it seems clear that nobody has a special right to this sink over anyone else. Now that the atmosphere as a GHG sink is a scarce good, regulation is in order. How ought the remaining carbon budget be allotted? Baatz argues that the principle of equality supports "equal per capita emissions rights" (EPCER). According to this principle, the calculation of an individual's "fair share" of emission rights is simple: "the total amount of available emissions is divided by the world population" (5). In 2014, sustainable equal per capita emissions were (arguably) 2–3t of CO_2; in principle, then, individuals have a moral obligation to reduce their emissions to this level. However, Baatz thinks an individual's emissions reduction duty is weaker *in practice* because of the intuitive moral principle that individuals are entitled to their *subsistence emissions*: those emissions that are required to maintain a minimally decent life. This qualification means that an individual's mitigation duty will be highly context-dependent and depends on the contested details of a "minimally decent life." Although one might think, as Baatz does, that in "carbon-dependent structures, subsistence emissions can be rather high" (10), nevertheless he argues that the demand to take only one's fair share remains serious.

For discussion: Do you agree with the fair shares approach? What do you think your fair share is, and how would you go about limiting your consumption to achieve it? Reflect on whether it is reasonable for people to give different answers to this question depending on where they live and the other advantages in life they enjoy.

Ty Raterman. (2012). Bearing the weight of the world: On the extent of an individual's environmental responsibility. *Environmental Values* 21: 417–436.

Early on in philosophical discussions of individual climate responsibilities, the details of the duty in question were left mostly unspecified. In large part, this is due to the dichotomous nature of this debate: *either* the individual has an obligation to adopt a sustainable lifestyle, *or* their duties lie entirely elsewhere. In the first half of this paper, philosopher Ty Raterman presents his own reflections on why he is convinced *some* individual environmental responsibility must exist. Then, he turns to his primary question of interest: if such a duty exists, what does it look like? Raterman argues against the Kantian answer: that one has the moral obligation to act in a sustainable manner regardless of what others do. Instead, he insists on a middle way between an individual obligation to live sustainably and no individual responsibilities whatsoever. Specifying the details of this duty may be impossible, but he offers a number of analogies to illustrate the central upshot of the paper: like in many other cases in life, "one is not expected to be as fully devoted to [environmentally friendly actions] as one could possibly be, but one is plainly not doing enough if one stays entirely within the realm of comfort and convenience" (433).

For discussion: What do you think of the upshot of Raterman's paper? Do you find the analogies he provides helpful for understanding the extent of individual climate responsibilities? What might be one possible critique of his view?

John Broome. (2016). A reply to my critics. *Midwest Studies in Philosophy* 40(1): 158–171.

In his 2012 book *Climate Matters*, philosopher John Broome argued that individuals had a moral obligation to reduce their emissions to net-zero. Here, he defends this view from several critics. His argument is relatively simple, at least on its face: emitting greenhouse gas is an injustice done to other people because it exposes them to a risk of harm, and this is wrong even if you did them no actual harm. The risk of harm you impose in the climate change case is large because of the great many people you risk harming, which amounts to a serious injustice. Moreover, because one has the option to offset any emissions they can't prevent from the outset, one has the obligation to offset in this way. In fact, Broome thinks offsetting is relatively easy. His critics challenge Broome on multiple fronts, which he addresses in turn.

For discussion: What do you think about Broome's argument for a moral obligation to reduce personal emissions to zero? Do you agree with the role he allows for offsetting in meeting this obligation? Of the objections he considers, which one do you think is the most serious?

Keith Hyams & Tina Fawcett. (2013). The ethics of carbon offsetting. *WIREs: Climate Change* 4, 91–98.

In cases where individuals or organizations feel compelled to engage in a high-emitting activity, but still wish to reduce their contributions to climate change, they may choose to "offset" their carbon emissions through investing in forestry or carbon capture and storage projects. If climate change is caused by a net increase of carbon emissions, the thought goes that one can eliminate one's own contribution to the problem by balancing out carbon added to the atmosphere with carbon withdrawn. In this survey article, political philosopher Keith Hyams and environmental researcher Tina Fawcett identify two basic types of criticisms leveled against offsetting: (1) that in fact offsetting programs will not accomplish what they set out to achieve, and (2) that carbon emitting is wrong in itself, so that offsetting is nothing more than a morally corrupt license to justify ongoing bad behavior. The first type of critique questions the scientific legitimacy of offsetting programs: Will an equivalent amount of carbon to what was added *actually* be removed from the atmosphere? There are good reasons to think, in many cases at least, the answer is *no*. The second type of critique tends to reject the consequentialist flavor of offsetting. Such a view insists that emitting carbon is wrong, full stop, and so can't be "offset." The practical implications of this dispute are potentially huge. If the criticisms of offsetting can be successfully rebutted, the content of the individual duty to reduce emissions may change significantly (Broome, 2012), especially since off-sets are currently still relatively cheap. If the criticisms cannot be overcome, then the demands on individuals may quickly become extreme as the climate budget continues to dwindle.

For discussion: What is your attitude to offsets? Are they simply a convenient and morally unproblematic tool? Or do they amount to selling "environmental indulgences"?

Dialogue 4: Climate Justice

Simon Caney. (2010). Cosmopolitan justice, responsibility, and global climate change. In S. Gardiner, S. Caney, D. Jamieson, & H. Shue (Eds.), *Climate Ethics: Essential Readings* (pp. 122–145). Oxford: Oxford University Press.

Once one accepts that global emissions must be reduced as a means of mitigating dangerous climate change, a question arises: how should the burdens of global decarbonization and/or adaptation be distributed? Who should "foot the bill"? On the face of it, there seems to be an obvious answer. Climate change is not merely a force of nature but is caused by humans. Moreover, not everybody seems equally culpable: the United States, for example, has produced the largest amount of historical emissions. Should the United States, therefore, bear the brunt of the burdens associated with fixing the climate problem? The *polluter pays principle*, already a significant principle in environmental law, seems to reinforce this idea. However, philosopher Simon Caney shows how polluter pays might take several incompatible forms, and argues that it faces interpretive and practical problems when applied to responsibility for climate change. He therefore suggests that the principle must be supplemented in various ways. Ultimately, he endorses a hybrid account, which gives a significant role to an ability to pay principle, as well as relying to some extent on polluter pays, but is based in different considerations. Caney outlines the similarities and differences between his revised responsibility

model and the concept of "common but differentiated responsibility" expressed in the 1992 Rio Declaration.

For discussion: Do you agree with the polluter pays principle as applied to climate change? Would an ability to pay
approach be better? What are the implications of focusing on
one rather than the other?

Henry Shue. (1993). Subsistence emissions and luxury emissions.
Law & Policy 15(1): 39–60.

In this classic early work on climate justice, philosopher Henry
Shue takes a look at the questions of fairness that arise in the
context of international climate negotiations. He identifies
four central questions: (1) What is a fair distribution of the
burden of mitigating climate harms? (2) What is a fair distribution of the burden of adapting to unavertable climate
changes? (3) What background conditions must be met before
climate negotiations can be executed fairly? (4) What is a fair
distribution of the remaining greenhouse gas emissions com-
patible with preventing global warming? He then argues that
any attempt to answer these questions will turn significantly on
whether one adopts fault-based or no-fault principles. Next,
he claims that there is a fundamental moral difference between
essential and non-essential emissions, which he calls the difference between "subsistence" and "luxury" emissions. Any
approach to climate justice that does not recognize this critical
distinction – such as any "homogenizing" account that considers all greenhouse gas emissions on a par regardless of their
source – is inadequate. If not all emissions are, morally speaking, equivalent, then equal-per-capita and other approaches to
fair emissions allocation insensitive to the *origins* of emissions
will be unacceptable. Ostensibly, this significantly complicates
fair climate negotiations.

For discussion: What do you think of the idea of distinguish-
ing subsistence and luxury emissions? How would you do that
within your own life and culture? Do you think one should

have different accounts of subsistence emissions and luxury emissions in different countries?

Darrel Moellendorf. (2014). Danger, poverty, and human dignity. In *The Moral Challenge of Dangerous Climate Change: Values, Poverty, Policy*. Cambridge: Cambridge University Press.

In the first chapter of his 2014 book, philosopher Darrel Moellendorf frames the climate problem in terms of the central objective of the United Nations Framework Convention on Climate Change (UNFCCC), avoiding "dangerous anthropogenic interference" in the climate system (9). He claims that this objective requires an account of dangerous climate change. He proposes that "judging something to be dangerous relies on there being reason to avoid it" (10) and that we understand the concept of "danger" as "too risky in light of the available alternatives" (10). He then argues that in the climate context this idea should be cashed out in terms of global poverty – "whether poverty eradication is delayed by either climate change or climate change policy is fundamental when identifying either as dangerous" (10). He takes this to imply that "any energy policy that prolongs global poverty is unreasonable" (22), and that the need for the poor to have access to cheap energy places constraints on how quickly we should decarbonize. More formally, he proposes an Anti-Poverty Principle (APP): policies and institutions should not impose any costs of climate change or climate change policy (such as mitigation and adaptation) on the global poor, of the present or future generations, when those costs make the prospects for poverty eradication worse than they would be absent them, if there are alternative policies that would prevent the poor from assuming those costs (22). Moellendorf highlights a potential conflict between poverty eradication and aggressive climate action and challenges the reader to consider how states ought to navigate this tension if it arises.

For discussion: Reflect on this principle and its potential implications. In particular, while initially it may seem straightforward, notice that on closer examination it appears to be surprisingly uncompromising. For example, if it were the case that keeping warming below 1.5°C delayed poverty eradication *even slightly*, then the principle dictates that justice requires a warmer world. Do you agree? More widely, do you believe that eliminating poverty trumps all other moral objectives? What would the world look like if it did?

Nancy Tuana. (2019). Climate apartheid: The forgetting of race in the Anthropocene. *Critical Philosophy of Race* 7(1): 1–31.

In this paper, philosopher Nancy Tuana argues that approaches to climate justice that merely analyze differential impacts, or the unfair distribution of benefits and burdens, are inadequate. Instead, she maintains that climate justice requires *genealogical sensibilities*: a rich understanding of the histories and lineages of the deep incorporation of racism and environmental exploitation. Attention to these histories will reveal both why color-blind climate policy is not a morally permissible option, and why it has been the dominant norm up to this point. She analyzes the concept of *climate apartheid* and explains why it is important. Like historical apartheid in South Africa, climate injustice should not be understood merely in terms of distributional inequality, since this framing masks the way differential impacts occur due to "deeply held systematic beliefs and dispositions regarding racial superiority" that are "supported by various social institutions" (5). So long as these root causes are left alone, systematic subordination and climate injustice will continue. Accordingly, Tuana urges us to scrutinize mitigation and adaptation strategies, always asking "for what and to whom" are they intended (8)? She illustrates this idea of climate justice with a genealogical sensibility through analysis of The Great Wall of Lagos, the ongoing environmental racism of the coal industry, and the conditions that allow for contemporary slavery in Brazil.

For discussion: Reflecting on this paper, you might ask yourself what does a genealogically informed climate justice movement look like? Can you think of examples familiar to you where the approach would make a difference? Should a genealogical approach inform international negotiations? If so, how?

Sara Mersha. (2018). Black lives and climate justice: Courage and power in defending communities and Mother Earth. *Third World Quarterly* 39(7): 1421-1434.

In this paper, activist Sara Mersha shares examples of leadership in Black communities struggling for climate justice around the world. In so doing, Mersha crystalizes how climate change is a racial justice issue and why it is galvanizing black and indigenous justice movements. Throughout, she scrutinizes the policies of high emitters and their aid. Whether its endorsement of the carbon offset program REDD (Reducing Emissions from Deforestation and Forest Degradation) or support for Green Revolution technologies, she argues that the mainstream policies by transnational corporations and governments consistently ignore grassroots climate advocacy. In addition, external aid designed to alleviate climate impacts frequently does not seek out or heed the wishes of local leaders, undermining the self-determination of vulnerable communities of color and stalling their adaptation strategies. This essay is a reminder of the grassroots climate activism thriving in the communities most vulnerable to climate impacts. In addition, it highlights how the policies for which these communities explicitly advocate are often incongruent with the way international forces respond.

For discussion: The reader should consider how the unjust background Mersha identifies should inform international climate negotiations and policy. What differences might it make to how both the process and outcomes of achieving climate justice should be understood?

K.P. Whyte. (2016). Indigenous peoples, climate change loss and damage, and the responsibility of settler states. Available at SSRN: https://ssrn.com/abstract=2770085 or http://dx.doi.org/10.2139/ssrn.2770085

> Adaptation and loss & damage are intimately linked. If nations can be shown to be morally responsible for climate-induced loss and damage to marginalized peoples, then a strong moral case can be made that they would also be politically responsible for abating them. In this paper, philosopher Kyle Powys Whyte argues that settler states are morally responsible for the losses and damages to indigenous peoples in at least two ways. First, settler states have *impending* responsibility for loss and damage: they are disproportionately responsible for the anthropogenic factors leading to climate change impacts (2). Second, settler states have *pending* responsibility: they are responsible for ongoing problems in political relations with indigenous peoples that deny them the capacity to effectively adapt to environmental change (3). Therefore, just climate action requires that settler states engage in radical political reconciliation with indigenous peoples. Only this progress, Whyte argues, could enable adaptation consistent with indigenous self-determination. The settler states' political responsibility in this way is immense: no just climate treaty can exist without the complete transformation of political relations, both at home and on the international stage.
>
> *For discussion:* Do you think this is a target for which international climate negotiations should aim? Is climate justice possible if it doesn't? What might be some reasons to adopt a different goal, and how might this change negotiations?

Dialogue 5: A Big Technological Fix?

Henry Shue. (2018). Climate dreaming: Negative emissions, risk transfer, and irreversibility. *Journal of Human Rights and the Environment* 8(2): 203–216.

IPCC models make clear that aggressive mitigation will be required to keep global temperature rise within 2 degrees, but it is easy to miss that even these scenarios rely heavily on negative emissions technologies (NETs). According to the philosopher Henry Shue, this is little better than adopting an attitude of "something will come up," which amounts to a highly unjust gamble. In addition to feasibility concerns, such reliance on NETs brings unacceptable moral costs in several ways. For one, bio-energy combined with carbon capture and storage (BECCS) – one of the most exciting NET prospects – has high land-use intensity. As a result, *just* large-scale deployment would require practically "surgical precision" in public policy. On top of this, if NETs prove unfeasible or unjust at scale, the moral costs will be felt by the already marginalized. Finally, even if the gamble paid off, this success would not justify less aggressive mitigation now as the IPCC models tacitly assume. Considering all this, Shue asserts that the only morally justifiable thing now is ambitious mitigation immediately. Any case for climate policy ethically dependent on NETs is a product of either self-serving moral corruption or bald lying about their prospects, and should be fought against in every effective way.

For discussion: The reader should consider whether they agree in this strong censure of NETs, or if they think NETs should play a role in current climate policy. If they should, how should their use affect mitigation policy and temperature targets?

Kent A. Peacock. (2021). As much as possible, as soon as possible: Getting negative about emissions. *Ethics, Policy & Environment*. Available at: www.tandfonline.com/doi/epub/10.1080/21550085.

Ethical analysis of carbon dioxide removal strategies (or negative emissions technologies, NETs) has lagged behind ethical scrutiny of mitigation, solar radiation management, and even adaptation. Here, philosopher Kent Peacock attempts to get other philosophers interested in climate to attend to this pressing ethical issue. After summarizing the scientific and political state of NETs, Kent Peacock presents the *prima facie* ethical case for their large-scale deployment *as soon as possible*. Quite simply, IPCC modeling shows there is little hope of holding warming to tolerable levels unless a substantial portion of existing atmospheric CO_2 is removed, and soon. NETs have the potential to fill this role and carry no principled downsides so long as they are not considered a *replacement* for aggressive mitigation. Finally, given that scaling up CDR could take several decades, we should start in this direction *now*. After warding off some common objections, Peacock concludes this piece by offering a policy proposal for facilitating the global draw down of CO_2 levels while guarding against the lure of mitigation deferral. In contrast to Shue's mitigation-only ethos, Peacock seems to adopt what might be called an "all hands on deck" approach.

For discussion: Does Peacock's policy risk deemphasizing mitigation or other costs? If it does, how might these be minimized? Are the risks worth it?

Doreen E. Martinez. (2014). The right to be free of fear: Indigeneity and the United Nations. *Wicazo Sa Review* 29(2): 63–87.

One constant risk embedded in climate responses – whether mitigation, adaptation, or geoengineering – is the exacerbation of environmental harms against already marginalized peoples. In this paper, sociologist Doreen Martinez highlights how climate responses can frequently come into conflict with the essential value of cultural self-determination, thereby exacerbating what she calls *climate colonialism*. She examines various United Nations' practices as exemplary of this problem. Specifically, she condemns "Hopenhagen" as marketing a hope of a climate future beneficial to all, insensitive to the catastrophic environmental impacts already felt by marginalized communities worldwide (72–74). The invisibility of indigenous communities further manifests in the focus on food production rather than food security, which risks perpetuation of water depletion and land grabs (77–78). Even the very conceptualism of global warming as climate change obscures the distinctly colonial act of forcibly modifying the climate (79). Martinez' powerful illustration of the exclusionary international arena raises difficult questions for the prospect of politically legitimate geoengineering, even if deployed by some form of global governance.

For discussion: Could land-intensive technologies like afforestation and BECCS be deployed equitably in this context? Should Solar Radiation Management itself be seen as one of the most literal forms of climate colonialism?

Stephen M. Gardiner. (2010). Is "arming the future" with geoengineering really the lesser evil? Some doubts about the ethics of intentionally manipulating the climate system. In Gardiner, S.M., Caney, S., Jamieson, D., & Shue, H. (Eds.). *Climate Ethics: Essential Readings.* Oxford: Oxford University Press.

The term "geoengineering" lacks a precise definition, but is widely held to imply the intentional manipulation of the environment on a global scale. Some argue our response to climate change has been so dismal so far that we should start preparing for the nightmare scenario where we are forced to choose between attempting geoengineering and allowing a catastrophe to occur. In such a scenario, the claim goes, geoengineering should be chosen as "the lesser evil"; and, given this, we should start doing serious research now on how best to geoengineer, so that we can "arm the future" with the right technology. Gardiner's main aim in this paper is to outline some of the ethical issues which complicate this argument for geoengineering. As a secondary matter, he argues for three more specific conclusions. First, the Arm the Future Argument assumes much that is contentious and is overly narrow in its conclusions. Second, the Argument obscures much of what is at stake in the ethics of geoengineering, including what it means to call something an "evil," and whether doing evil has further moral implications. Third, since the Argument arises in a troubling context – climate change is a perfect moral storm – its role in the debate should be viewed with suspicion.

For discussion: What do you think of the arm the future argument? Should stratospheric sulfate injection be seen as a lesser evil? Which sense of "evil" would that be? What are the implications?

Joshua Horton & David Keith. (2016). Solar geoengineering and obligations to the global poor. In C.J. Preston (Ed.), *Climate Justice and Geoengineering: Ethics and Politics in the Atmospheric Anthropocene*. New York: Roman & Littlefield.

At this point, it is widely acknowledged that climate-induced ecosystem stresses will disproportionately impact the global poor, who are least responsible for the impending climate crisis. In this paper, political scientist Joshua Horton and physicist David Keith argue that this asymmetry of climate impacts has crucial implications for the ethics of solar geoengineering. They take it as straightforward that requirements of justice are violated when an activity benefits wealthy countries at the expense of poorer ones. In such cases, the rich countries have a moral obligation to take steps to reduce the harms falling on vulnerable nations. Horton & Keith contend that if obligations to the global poor include not only the mitigation of future harms but also reducing climate harms in the near term – a reasonable assumption – then this duty cannot be fulfilled through mere mitigation. Rather, historical emitters have a duty to invest in adaptation or SRM research because these strategies have the potential to substantially reduce climate harms in the short term. Unlike adaptation, however, SRM is global in scale and can be executed at relatively low expense. Ultimately, given the evidence that SRM could significantly reduce global temperatures and limit climate impacts that would be felt most in the developing world, Horton & Keith contend that research on SRM is a moral imperative.

For discussion: Do you agree? Do you think it is likely that SRM will be deployed in such a way as to benefit the global poor? What considerations of justice would arise?

Marion Hourdequin. (2019). Climate change, climate engineering, and the "global poor": What does justice require? *Ethics, Policy & Environment* 21(3): 270–288. Also in S.M. Gardiner, C. McKinnon, and A. Fragnière (Eds.), *The Ethics of "Geoengineering" the Global Climate: Justice, Legitimacy and Governance*. Routledge.

Philosopher Marion Hourdequin takes on Horton and Keith's argument that research into solar radiation management (SRM) is morally obligatory because the resulting knowledge has the potential to benefit everyone, and in particular the global poor. While such arguments emphasize distributive and consequentialist concerns, Hourdequin maintains that they overlook procedural and recognitional justice, and thus relegate to the background questions of how SRM research should be governed. By contrast, she argues for a multidimensional approach to geoengineering justice. To overcome worries such as paternalism and parochialism, Hourdequin advocates for what she calls institutional participatory parity. Among other things, this entails that rather than simply claiming to act on behalf of the global poor, institutions promoting SRM research and governance have an obligation to involve stakeholders from disadvantaged communities from the outset.

For discussion: Which of Hourdequin's points do you find most interesting? How might Horton and Keith respond? What implications does this debate have for geoengineering policy, climate science more generally, and for your own view?

David R. Morrow. (2014). Starting a flood to stop a fire? Some moral constraints on Solar Radiation Management. *Ethics, Policy & Environment* 17(2): 123–138.

The permissibility of Solar Radiation Management (SRM) research depends, in part, on whether the *deployment* of SRM could ever be morally permissible. Frequently, scientists who think SRM use could, at least eventually, be morally justified assume that (1) an engineered climate would be safer than one subject to runaway warming, and (2) if SRM were the only way to achieve this safer climate, then it is morally permissible. Philosopher David Morrow challenges the second premise of this argument by invoking two well-established moral principles: the Doctrine of Doing and Allowing and the Doctrine of Double Effect. He explains and motivates these principles through a thought-experiment: a city opening a dam in order to start a flood that will extinguish a fire they were partly responsible for causing. Both principles, Morrow argues, set the standard for permissible SRM use higher than scientists usually suggest. He concedes that SRM may still be justified if the benefits of an engineered climate *vastly* exceed the costs of allowing climate change to accelerate, but he still insists that SRM can be justified only if the risks it creates are *much less* than the expected benefits. Ultimately, Morrow argues for a middle way between staunch SRM supporters and some of their critics. On the one hand, SRM may very well violate important moral constraints, and so passing a simple cost/benefit analysis or risk profile analysis does not justify deployment alone. On the other, the threat of climate change is great enough that SRM may eventually be needed to avoid catastrophe. In such a case, it is reasonable to think that the moral agent can justifiably forgo even otherwise prohibitive moral doctrines. To insist otherwise would be to accept a staggering ethical precept: *Fiat Justitia, ruat caelum* – "Let justice be done, though the heavens fall." Morrow concludes that the ethics of SRM research cannot be settled on paper but must be given a fair hearing in the public arena, with the moral reasons on both sides acknowledged and weighed.

For discussion: Would you start a flood to stop a fire? Reflecting on this paper, consider whether you agree that more than the consequences matter, and how strong concern for intention should be. What does your answer mean for the ethics of geoengineering?

Cristina Yumie Aoki Inoue. (2018). Worlding the study of global environmental politics in the Anthropocene: Indigenous voices from the Amazon. *Global Environmental Politics* 18(4): 25–42.

Increasingly across the academic disciplines, attention is being paid to how individuals' worldviews are dependent on their social and historical context. Here, international relations scholar Cristina Yumie Aoki Inoue argues this perspectivism undergirds the political importance of *worlding*: of critically situating one's theories and concepts as constitutive of one's own world and stretching one's views in time and space to uncover what is hidden by those concepts and assumptions. In so doing, one recognizes "that ours is but one among many worlds" (27). Worlding global environmental politics involves critically situating one's own theories and engaging in dialogue with other worldviews through Creative Listening and Speaking (CLS). Listening effectively to other ways of knowing requires approaching differing perspectives from a place of epistemic and ontological parity, and thereby changing relationships into mutuality. Inoue calls this place of discursive agency a *third space*. She closes her article by trying to create a third space to listen to Davi Kopenawa, a shaman of the South American Yanomami tribe. Perhaps unsurprisingly, she finds his perspective differs dramatically from dominant environmental framings in the Global North.

For discussion: Reflecting on Inuoue's paper, the reader should consider the prospects of worlding the global politics of geoengineering. Can it be done? In its absence, can global climate intervention be morally justified?

Dialogue 6: The Future

Derek Parfit. (1983a). Energy policy and the further future: The social discount rate. In D. Maclean & P. Brown (Eds.), *Energy and the Future* (pp. 31–37). Totowa, NJ: Rowman & Littlefield.

> Philosopher Derek Parfit critiques the concept of "economic discounting" and the role it plays in economic analysis. According to the "social discount rate" (SDR), all future costs and benefits may be "discounted" at some rate n per year. After surveying and scrutinizing various justifications that have been offered to defend the SDR, Parfit argues that they either provide no reason to discount, or that they justify discounting only in some situations. He concludes that the SDR, as it has typically been articulated and used by economists, is indefensible. Ultimately, mere difference in time is morally neutral. Parfit's conclusion has significant implications for all intergenerational social problems, and problematizes mainstream economic models of climate change in particular.
>
> *For discussion*: Which justification of the SDR do you consider weakest, and which strongest? What kind of moral judgments (if any) are bound up in your reasons?

Derek Parfit. (1983b). Energy policy and the further future: The identity problem. In D. Maclean & P. Brown (Eds.), *Energy and the Future* (pp. 166–179). Totowa, NJ: Rowman & Littlefield.

> Philosopher Derek Parfit assesses the assumption that contemporary policy decisions harm people in the distant future by making them better or worse off. He argues that, whereas contemporary acts and their immediate effects have definite victims or beneficiaries in persons who already exist, policy choices with very long-term effects affect the identity of those who will exist in the far future. According to Parfit, this fact problematizes many intuitive accounts of intergenerational obligation. In particular, for many bad environmental policies

the ostensible "victims" in the future cannot be said to have been made "worse off" or harmed. Rather than reject intergenerational obligations, Parfit proposes an alternative explanation of intergenerational duties that does not rely on current people making particular future people worse off than they would have been otherwise.

For discussion: Are you troubled by the nonidentity problem? Should it make a difference to how we understand our intergenerational climate duties?

Lukas H. Meyer & Dominic Roser. (2009). Enough for the future. In A. Gosseries & L. Meyer (Eds.), *Intergenerational Justice* (pp. 219–248). Oxford: Oxford University Press.

Philosophers Lukas Meyer and Dominic Roser defend one vision of what it means to do right by future people. They begin by differentiating sufficientarian and egalitarian conceptions of justice: sufficientarianism holds that justice consists in every person – present and future – meeting a specific threshold of well-being, while egalitarianism holds that relative differences in well-being are intrinsically unjust. Next, Meyer and Roser offer two main reasons to accept intergenerational sufficientarianism. First, a sufficientarian conception of justice can accommodate the non-identity problem more powerfully than egalitarianism; second, sufficientarianism explains why current people must protect the interests of future people while avoiding several problems faced by egalitarianism. This paper summarizes multiple significant debates in the intergenerational climate justice and defends an important line of argumentation in response to them.

For discussion: In light of this helpful survey, some may want to dive into some intergenerational ethical theorizing of their own. What notion of harm, if any, seems best (e.g., Parfit's or Meyer and Roser's)? What is the best way of conceptualizing our duties to future generations? What implications might your view on this take regarding our climate policies today?

Rivka Weinberg. (2008). Identifying and dissolving the non-identity problem. *Philosophical Studies* 137: 3–18.

> One common response to the non-identity problem is to supply a notion of harming future people that does not require making them worse off than they would otherwise be. Philosopher Rivka Weinberg takes a different tack by attempting to solve the problem on its own narrow person-affecting terms. Her solution is really a dissolution. Weinberg argues that the non-identity problem only poses a problem at all when *merely possible* people (e.g., completely hypothetical people who will never actually exist) are deemed morally relevant. She contends they are not morally relevant, and so the thrust of the paradox falters.
>
> *For discussion*: What do you think of Weinberg's argument? In particular, you might want to consider whether you believe that future people have an interest in existence itself (13–16). Do you agree with Weinberg's instinct that we should refuse to be forced into the dialectic entailed by the non-identity problem? (15)

Dale Jamieson. (2009). Climate Change, responsibility, and justice. *Science and Engineering Ethics* 16, 431–445.

> Philosopher Dale Jamieson argues that everyday morality cannot appropriately categorize climate change as a moral wrong that makes strong demands on persons and states. Paradigm cases of moral wrongs almost always involve unambiguous wrongdoers, discrete victims, and easily traceable harm. He provides one example of a paradigm moral problem – that of Jack intentionally stealing Jill's bike – to show how climate change lacks its clear wrong-making features. Accordingly, Jamieson argues that an entirely new sense of moral responsibility is needed. He begins this task but showing how climate change threatens another value – respect for nature – that is regularly overlooked.

For discussion: If Jamieson is right that current moral concepts are inadequate for grasping the severity of the climate problem, this may explain why robust climate action has not yet happened. Is he? If yes, then how might we need to revise our moral concepts? Will we be able to do so in time?

Stephen M. Gardiner. (2011b). Is no one responsible for global environmental tragedy? Climate change as a challenge to our ethical concepts. In D.G. Arnold (Ed.), *The Ethics of Global Climate Change* (pp. 38–59). Cambridge University Press.

Philosopher Stephen Gardiner resists Jamieson's view that climate change lacks the features of a paradigm moral problem and demands conceptual revision. Gardiner argues that everyday morality *does* provide many of the tools to understand why climate change is a moral wrong and offers an alternative to the Jack and Jill example to demonstrate this point. Ultimately, he claims that one central problem climate change poses is a failure of delegation: given that political institutions have failed to address the challenge, individuals must fight to create more just institutions. Neglecting to meet this demand is not a new problem, but is nonetheless jarring: it restores to individuals demanding social burdens to which we have become unaccustomed. While Gardiner's focus on delegation may be less revisionary than Jamieson's, he still concludes that the moral and political challenge we face is severe.

For discussion: Considering the exchange between Gardiner and Jamieson, do you think ongoing climate inaction is more a product of lacking the proper moral concepts, or simply a more well-trodden case of absent political will? What impact does your answer have for the approach you'd propose for overcoming political inertia on climate change?

Allen Thompson. (2010). Radical hope for living well in a warmer world. *Journal of Agricultural and Environmental Ethics* 23(1–2): 43–55.

In normal cases, hope has a clear object. One may hope for an end to war, a sickness cured, or love returned from another. In cases of hoping for a better future amid severe cultural change, however, the object of hope may be obscured. In such a situation, Jonathan Lear suggests radical hope is needed: "the hope for revival … for coming back to life in a form that is not yet intelligible" (Lear, 2008, p. 95). Philosopher Allen Thompson argues that radical hope will be needed to realize a future of human flourishing in a warmer world. Global warming will bring fundamental change. Present cultures of consumerism cannot coexist with it. Consumerism will not survive run-away warming, and avoiding run-away warming will require rapid decarbonization that is also inconsistent with consumerism. Moreover, global warming threatens to obliterate the autonomy of the natural world, rendering humanity fully responsible for Earth's future. This possibility is simply incongruent with the values of western environmentalism as it has come to be constructed. These changes, Thompson argues, ensure that the good life of the future must be very different than the good life today. Radical hope calls for facing this prospect with courage, finding the faith to act in pursuit of a better future one cannot even imagine.

For discussion: To achieve such a future, do you think human societies will need radical hope? Should radical hope be considered a human virtue, and if so how might this virtue be cultivated?

Full Bibliography

Ackerman, F., & Heinzerling, L. (2004). *Priceless: On Knowing the Price of Everything and the Value of Nothing*. New York: The New Press.

Agarwal, A., & Narain, S. (1991). *Global Warming in an Unequal World: A Case of Environmental Colonialism*. New Delhi: Centre for Science and Environment.

Aisch, G., Buchanan, L., Cox, A., & Quealy, K. (2017). Some colleges have more students from the top 1 percent than the bottom 60. Find yours. *The New York Times*. Available at: www.nytimes.com/interactive/2017/01/18/upshot/some-colleges-have-more-students-from-the-top-1-percent-than-the-bottom-60.html

Aldern, C.P. (2020, May 14). We're running out of time to flatten the curve – for climate change. *Grist*. Available at: https://grist.org/climate/flatten-the-curve-coronavirus-climate-emissions/

Anderson, K. (2012). Climate change going beyond dangerous: Brutal numbers and tenuous hope. *Development Dialogue* 61: 16–40.

Anderson, K., & Peters, G. (2016a). The promise of negative emissions – response. *Science* 354(6313): 714–715.

Anderson, K., & Peters, G. (2016b). The trouble with negative emissions. *Science* 354(6309): 182–183.

Archer, D., Kite, E., & Lusk, G. (2020). The ultimate cost of carbon. *Climatic Change* 162: 2069–2086.

Arrow, K. (1996). Intertemporal equity, discounting, and economic efficiency. In J. Bruce, H. Lee, & E. Haites (Eds.), *Climate Change 1995: Economic and Social Dimensions of Climate Change*. Cambridge: Cambridge University Press.

Aufrecht, M. (2011). Climate change and structural emissions: Moral obligations at the individual level. *International Journal of Applied Philosophy* 25: 201–213.

Baatz, C. (2014). Climate change and individual duties to reduce GHG emissions. *Ethics, Policy & Environment* 17(1): 1–19.

Baer, P. (2002). Equity, greenhouse gas emissions, and global common resources. In S. Schneider (Ed.), *Climate Change Policy: A Survey*. Washington, DC: Island Press.

Baer, P. (2010). Greenhouse development rights: A framework for climate protection that is 'more fair' than Equal Per Capita Emissions Rights. In S.

Gardiner, S. Caney, D. Jamieson, & H. Shue (Eds.), *Climate Ethics: Essential Readings* (pp. 215–230). Oxford: Oxford University Press.

Baer, P., Harte, J., Haya, B., Herzog, A.V., Holdren, J., Hultman, N.E., Kammen, D.M., Norgaard, R.B., & Raymond, L. (2000). Equity and greenhouse gas responsibility. *Science* 289(5488): 2287.

Barnes, E. (2016). *The Minority Body: A Theory of Disability.* Oxford: Oxford University Press.

Baragwamath, K., & Bayi, E. (2020). Collective property rights reduce deforestation in the Brazilian Amazon. *Proceedings of the National Academy of Sciences* 117.

Barnett, Z. (2020). Why you should vote to change the outcome. *Philosophy & Public Affairs* 48(4): 422–446.

Barrett, S. (1997). The strategy of trade sanctions in international environmental agreements. *Resource and Energy Economics* 19(4): 345–361.

Bar-Yam, Y., Lagi, M., & Bar-Yam, Y. (2015). South African riots: Repercussion of the global food crisis and US drought. In P. Fellman, Y. Bar-Yam, and A. Minai (Eds.), *Conflict and Complexity* (pp. 261–267). New York: Springer.

Bastin, J., Finegold, Y., Garcia, C., Mollicone, D., Revende, M., Routh, D., Zohner, C.M., & Crowther, T.W. (2019). The global tree restoration potential. *Science* 365(6448): 76–79.

Bathiany, S., Dakos, V., & Scheffer, M. (2018). Climate models predict increasing temperature variability in poor countries. *Science Advances* 4(5): 1–10.

BBC. (2009, December 19). Copenhagen deal reaction in quotes. Available at: http://news.bbc.co.uk/2/hi/science/nature/8421910.stm

Beckerman, W., & Hepburn, C. (2007). Ethics of the discount rate in the stern review on the economics of climate change. *World Economics* 8(1): 187–210.

Beitz, C.R. (1999). *Political Theory and International Relations: Revised Edition.* Princeton, NJ: Princeton University Press.

Bendik-Keymer, J. (2012). The sixth mass extinction is caused by us. In A. Thompson and J. Bendik-Keymer (Eds.), *Ethical Adaptation to Climate Change: Human Virtues of the Future.* Cambridge, MA: MIT Press.

Blomfield, M. (2015). Geoengineering in a climate of uncertainty. In J. Moss (Ed.), *Climate Change and Justice* (pp. 39–58). Cambridge: Cambridge University Press.

Bokuluch, P. (2013, November 18). Gish gallop. *(Pseudo-)Science Blog.* Available at: https://denialism.wordpress.com/page/3/ (Accessed: 6 October, 2021).

Booker, B. (2015). Rule consequentialism. *Stanford Encyclopedia of Philosophy.* Available at:

Boonin, D. (2014). *The Non-Identity Problem and the Ethics of Future People.* Oxford: Oxford University Press.

Boonin, D. (2021). Parfit and the non-identity problem. In S. Gardiner (Ed.), *The Oxford Handbook of Intergenerational Ethics.* Oxford: Oxford University Press.

Boyd, P.W., *et al.* (2007). Mesoscale iron enrichment experiments 1993–2005: Synthesis and future directions. *Science* 315(5812).

Broome, J. (2012). *Climate Matters: Ethics in A Warming World.* New York: W.W. Norton.

Broome, J. (2016). A reply to my critics. *Midwest Studies in Philosophy* 40(1): 158–171.

Broome, J. (2019). Against denialism. *The Monist* 102(1): 110–29.

Brown, D.A. (2012). *Climate Change Ethics: Navigating the Perfect Moral Storm.* New York: Routledge.

Buchanan, J.M. (1965). An economic theory of clubs. *Economica* 32(): 1–14.

Budolfson, J.M. (2019). The inefficacy objection to consequentialism and the problem with the expected consequences response. *Philosophical Studies* 176: 1711–1724.

Bullard, R.D. (2000). *Dumping in Dixie: Race, Class, and Environmental Quality.* Boulder, CO: Westview Press.

Bullard, R.D. (2001). Environmental justice in the 21st century: Race still matters. *Phylon* 49(3/4): 151–171.

Burrell, A.L., Evans, J.P., & De Kauwe, M.G. (2020). Anthropogenic climate change has driven over 5 million km² of drylands towards desertification. *Nature communications* 11(2): 1–11.

Callicott, J.B. (1991). The wilderness idea revisited: The sustainable development alternative. *Environmental Professional* 13(3): 235–247.

Callicott, J.B., & Nelson, N.P. (1998). *The Great New Wilderness Debate.* Athens, GA: University of Georgia Press.

Callies, D.E. (2019). *Climate Engineering: A Normative Perspective.* New York: Lexington Books.

Caney, S. (2010). Cosmopolitan justice, responsibility, and global climate change. In S. Gardiner, S. Caney, D. Jamieson, & H. Shue (Eds.), *Climate Ethics: Essential Readings* (pp. 122–145). Oxford: Oxford University Press.

Caney, S. (2014). Two kinds of climate justice: Avoiding harm and sharing burdens. *The Journal of Political Philosophy* 22(2): 125–149.

Capisani, S. (2020). Territorial instability and the right to a livable locality. *Environmental Ethics* 42: 189–207.

Capstick, S., Khosla, R., Wang, S., et al. (2020). Bridging the gap – the role of equitable low-carbon lifestyles. In *Emissions Gap Report 2020.* UNEP. Available at: https://wedocs.unep.org/20.500.11822/34432

Carton, W. (2020). Carbon unicorns and fossil futures: Whose emission reduction pathways is the IPCC preforming? In J.P. Sapinksi, H.J. Buck, & A. Malm (Eds.). *Has It Come to This? The Promises and Perils of Geoengineering on the Brink.* New Brunswick, NJ: Rutgers University Press.

Carr, W., & Preston, C.J. (2017). Skewed vulnerabilities and moral corruption in global perspectives on climate engineering. *Environmental Values* 26(6): 757–777.

Center for Disaster Philanthropy. (2020, December 7). 2020 North American wildfire season. Available at: https://disasterphilanthropy.org/disaster/2020-california-wildfires/ (Accessed: October 8, 2021).

CDM-Watch. (n.d). *Home.* Available at: www.cdm-watch.org/ (Accessed October 5, 2021).

Chancel, L., & Piketty, T. (2015). *Carbon and inequality: From Kyoto to Paris.* Paris: Paris School of Economics.

Chichilnisky, G. (1996). An axiomatic approach to sustainable development. *Social Choice and Welfare* 13(2): 231–257.

Choi-Schagrin, W. (2021, August 24). Wildfires are ravaging forests set aside to soak up greenhouse gases. *The New York Times.*

Cicerone, R. (2006). Geoengineering: Encouraging research and overseeing implementation. *Climactic Change* 77: 221–226. Available at: https://doi.org/10.1007/s10584-006-9102-x

Cohen, G.A. (2009). *Why Not Socialism?* Princeton, NJ: Princeton University Press.

Consoli, C. (2019). *Bioenergy and Carbon Capture and Storage*. Docklands, Australia: Global CCS Institute. Available at: www.globalccsinstitute.com/resources/publications-reports-research/bioenergy-and-carbon-capture-and-storage/

Constanza, R. d'Arge, R., de Groot, R., Farber, S., Grasso, M., Hannon, B., Limburg, K., Naeem, S., Paruelo, J., O'Neill, R.V., & Raskin, R. (1997). The value of the world's ecosystem services and natural capital. *Nature* 387: 253–260.

Cook, J., Oreskes, N., Doran, P.T., Anderegg, W.R., Verheggen, B., Maibach, E.W., Carlton, J.S., Lewandowsky, S., Skuce, A.G., Green, S.A., Nuccitelli, D., Jacobs, P., Richardson, M., Winkler, B., Painting, R., & Rice, K. (2016). Consensus on consensus: A synthesis of consensus estimates on human caused global–warming. *Environmental Research Letters* 11(4): 1–7.

Cordato, R.E. (2001). *The Polluter Pays Principle: A Proper Guide for Environmental Policy*. Washington, DC: Institute for Research on the Economics of Taxation.

Cowen, T., & Parfit, D. (1992). Against the social discount rate. In P. Laslett & J.S. Fishkin (Eds.), *Justice Between Age Groups and Generations* (pp. 141–161). New Haven, CT: Yale University Press.

Cripps, E. (2013). *Climate change and the moral agent*. Oxford: Oxford University Press.

Cronon, W. (1995, August 13). The trouble with wilderness. *The New York Times*. Available at: www.nytimes.com/1995/08/13/magazine/the-trouble-with-wilderness.html

Crutzen, P.J. (2002). Geology of mankind. *Nature* 415: 23.

Crutzen, P.J. (2006). Albedo enhancement by stratospheric sulfur injections: A contribution to resolve a policy dilemma? *Climatic change* 77(211): 211–219.

Cullity, G. (2019). Climate Harms. *The Monist* 201(1): 22–41.

Cuomo, C. (2011). Climate change, vulnerability, and responsibility. *Hypatia* 26(4): 690–714.

Cuomo, C. (2021). Respect for nature: Learning from Indigenous values. In J. Kawall (Ed.), *The Virtues of Sustainability*. Oxford: Oxford University Press.

den Elzen, M. (1999). *Report on the expert meeting on the Brazilian Proposal: Scientific aspects and data availability*. Available at: http://unfccc.int/resource/brazil/documents/mrep1999.pdf

Department of Defense (2019). *Report on Effects of a Changing Climate to the Department of Defense*. Available at: https://media.defense.gov/2019/Jan/29/2002084200/-1/-1/1/CLIMATE-CHANGE-REPORT-2019.PDF

Department of Defense. (2010). *Quadrennial Defense Review Report*. Available at: https://history.defense.gov/LinkClick.aspx?fileticket=tFG_5ZPMhBk%3d&tabid=9114&portalid=70&mid=20230

Descartes, R. (2008). *Meditations on First Philosophy: With Selections from the Objection and Replies*. M. Moriarty (Ed./Trans.). Oxford: Oxford University Press.

Desch, S.J., Smith, N., Groppi, C., Vargas, P., Jackson, R., Kalyaan, A., Nguyen, P., Probst, L., Rubin, M.E., Singleton, H., Spacek, A., Truitt, A., Zaw, P.P., & Hartnett, H.E. (2017). Arctic ice management. *Earth's Future* 5: 107–127.

Dimitrov, R., Hovi, J., Sprinz, D., Sælen, H., & Underdal, A. (2019). Institutional and environmental effectiveness: Will the Paris Agreement work? *WIREs: Climate Change* 10(4): 1–12.

Donner, S. (2007). Domain of the Gods: An editorial essay. *Climatic Change* 85(3): 231–236.

Douglas, H. (2000). Inductive risk and values in science. *Philosophy of Science* 67: 559–579.

Douglas, H. (2009). *Science, Policy, and the Value-Free Ideal.* Pittsburgh, PA: University of Pittsburgh Press.

Dout, C.C., & Obst, A.R. (Forthcoming). Individual responsibility and the ethics of hoping for a more just climate future. *Environmental Values.*

Doyle, J. (2011). Where has all the oil gone? BP branding and the discursive elimination of climate change risk. In N. Heffernan & D. Wragg (Eds.), *Culture, Environment & Eco-Politics.* Newcastle: Cambridge Scholars Press.

Dunlap, R.E. (2013). Climate change skepticism and denial: An introduction. *American Behavioral Scientist* 57(6): 691–698.

Elgin, B. (2020, December 9). These trees are not what they seem. *Bloomberg Green.* Available at: www.realclearenergy.org/2020/12/14/these_trees_are_not_what_they_seem_652588.html

Ellis, E.C. (2015). Too big for nature. In B.A. Minteer & S.J. Pyne (Eds.), *After Preservation: Saving American Nature in the Age of Humans* (pp. 24–31). Chicago, IL: University of Chicago Press.

Environmental Pollution Panel & President's Science Advisory Committee. (1965). *Restoring the quality of our environment.* Washington, DC: The White House.

EPA. (n.d.) Greenhouse gasses equivalences calculator – calculations and references. Available at: www.epa.gov/energy/greenhouse-gases-equivalencies-calculator-calculations-and-references (Accessed: October 5, 2021).

European Court of Auditors. (2021). *The Polluter Pays Principle: Inconsistent application across EU environmental policies and actions.* Available at: www.eca.europa.eu/Lists/ECADocuments/SR21_12/SR_polluter_pays_principle_EN.pdf

Fajardy, M., Köberle, A., Dowell, N.M., & Fantuzzi, A. (2019). BECCS deployment: A reality check. *Grantham Institute Briefing Paper* 28.

Farmer, M., & Chiquiza, J.P. (2020, September 11). The smoke is here. Hazardous air quality in Seattle as wildfires rage. *KUOW.* Available at: www.kuow.org/stories/photos-massive-plume-of-smoke-arrives-in-seattle

Fenston, J. (2019, July 16). D.C. averages a week of 100-degree days. Climate change could make that two months. *WAMU88.5.* Available at: https://wamu.org/story/19/07/16/d-c-averages-a-week-of-100-degree-days-climate-change-could-make-that-two-months/ (Accessed: October 6, 2021).

Fiala, A. (2017). Anarchism. *Stanford Encyclopedia of Philosophy.* Available at: https://plato.stanford.edu/archives/spr2018/entries/anarchism/ (Accessed: October 6, 2021).

Fialka, J. (2020, January 23). U.S. geoengineering research gets a life with $4 million from Congress. *Science.* Available at: www.science.org/content/article/us-geoengineering-research-gets-lift-4-million-congress

Field, L., Ivanova, D., Bhattacharyya, S., Mlaker, V., Sholtz, A., Decca, R., Manzara, A., Johnson, D., Christodoulou, E., Water, P., & Katuri, K. (2018). Increasing Arctic sea ice albedo using localized reversible geoengineering. *Earth's Future* 6: 882–901.

Flavelle, C. (2020, October 28). As climate disaster pile up, a radical proposal gains traction. *New York Times*.

Fleurbaey, M., Ferranna, M., Budolfson, M., Denning, F., Mintz-Woo, K., Socolow, R., Spears, D., & Zuber, S. (2019). The social cost of carbon: Valuing inequality, risk and population for climate policy. *The Monist* 102(1): 84–109.

Foot, P. (1984). Killing and letting die. In J.L. Garfield & P. Hennessey (Eds.), *Abortion: Moral and Legal Perspectives* (pp. 175–184). Amherst, MA: University of Massachusetts Press.

Fragnière, A. (2016). Climate change and individual duties. *WIREs: Climate Change* 7(6): 798–814.

Fragnière, A. (2018). How demanding Is our climate duty? An application of the no-harm principle to individual emissions. *Environmental Values* 27: 645–663.

Fragnière, A., & Gardiner, S.M. (2016). Why geoengineering is not "Plan B." In C. Preston (Ed.), *Climate Justice and Geoengineering* (pp. 15–31). New York: Rowman & Littlefield.

Fricker, M. (2007). *Epistemic Injustice*. Oxford: Oxford University Press.

Frank, R. (2005). *Microeconomics and Behavior* (6th ed.). Boston, MA: McGraw-Hill.

Friends of the Earth International. (2021). *Chasing Carbon Unicorns: The Deception of Markets and 'Net Zero'*. FOEI. Available at: www.corporateaccountability.org/wp-content/uploads/2021/02/FoEI-carbon-unicorn-report-ENG-lr.pdf

Frumhoff, P. (2014). Global warming fact: more than half of all industrial CO_2 pollution has been emitted since 1988. *Union of Concerned Scientists Blog*. Available at: https://blog.ucsusa.org/peter-frumhoff/global-warming-fact-co2-emissions-since-1988-764 (Accessed: October 6, 2021).

Frumkin, H., Bratman, G.N., Breslow, S.J., Cochran, B., Khan Jr, P.H., Lawler, J.J., Levin, P.S., Tandon, P.S., Varanasi, U., Wolf, K.L., & Wood, S.A. (2017). Nature contact and human health: A research agenda. *Environmental Health Perspectives* 125(7). Available at: https://ehp.niehs.nih.gov/doi/10.1289/EHP1663

Gambhir, A., & Tavoni, M. (2019). Direct Air Carbon Capture and Sequestration: How it works and how it could contribute to climate-change mitigation. *One Earth* 1(4): 405–409.

Gandhi, L. (1998). *Postcolonial Theory: A Critical Introduction*. New York: Columbia University Press.

Gandhi, L. (2019). *Postcolonial Theory: A Critical Introduction*. Columbia, MD: Columbia University Press.

Gardiner, S.M. (2001). The real tragedy of the commons. *Philosophy & Public Affairs* 30(4): 387–416.

Gardiner, S.M. (2004). Ethics and global climate change. *Ethics* 144: 555–600.

Gardiner, S.M. (2010). Is "arming the future" with geoengineering really the lesser evil? Some doubts about the ethics of intentionally manipulating the climate system. In S.M. Gardiner, S. Caney, D. Jamieson, & H. Shue (Eds.), *Climate Ethics: Essential Readings* (pp. 284–312). Oxford: Oxford University Press.

Gardiner, S.M. (2011a). *A Perfect Moral Storm: The Ethical Tragedy of Climate Change.* Oxford: Oxford University Press.

Gardiner, S.M. (2011b). Is no one responsible for global environmental tragedy? Climate change as a challenge to our ethical concepts. In D.G. Arnold (Ed.), *The Ethics of Global Climate Change* (pp. 38–59). Cambridge University Press.

Gardiner, S.M. (2011c). Some early ethics of geoengineering the climate: A commentary on the values of the Royal Society report. *Environmental Values* 20(2): 163–188.

Gardiner, S.M. (2012). Are we the scum of the earth? Climate change, geoengineering, and humanity's challenge. In A. Thompson & J. Bendik-Keymer (Eds.), *Ethical Adaptation to Climate Change: Human Virtues of the Future.* Cambridge, MA: MIT Press.

Gardiner, S.M. (2013a). Geoengineering and moral schizophrenia: What's the question? In W. Burns & A. Strauss (Eds.), *Climate Change Geoengineering: Legal, Political and Philosophical Perspectives* (pp. 11–38). Cambridge: Cambridge University Press.

Gardiner, S.M. (2013b). The desperation argument for geoengineering. *PS: Political Science and Politics* 46(1): 28–33.

Gardiner, S.M. (2013c). Why geoengineering is not a global public good, and why it is ethically misleading to frame it as one. *Climatic Change* 121: 513–525.

Gardiner, S.M. (2014). A call for a Global Constitutional Convention focused on Future Generations. *Ethics and International Affairs* 28(3): 299–315.

Gardiner, S.M. (2016a). Geoengineering: Ethical questions for deliberate climate manipulators. In S.M. Gardiner & A. Thompson (Eds.), *The Oxford Handbook of Environmental Ethics.* Oxford: Oxford University Press.

Gardiner, S.M. (2016b). In defense of climate ethics. In S.M. Gardiner & D.A. Weisbach, *Debating Climate Ethics.* Oxford: Oxford University Press.

Gardiner, S.M. (2017a). Climate ethics in a dark and dangerous time. *Ethics* 127: 430–465.

Gardiner, S.M. (2017b). The threat of intergenerational extortion: on the temptation to become the climate mafia, masquerading as an intergenerational Robin Hood. *Canadian Journal of Philosophy* 47(2–3): 368–394.

Gardiner, S.M. (2017c). Trump and climate justice. *The Philosophers' Magazine*, 78: 14–16.

Gardiner, S.M. (2017d). Accepting collective responsibility for the future. *Journal of Practical Ethics* 5(1). Available at: www.jpe.ox.ac.uk/papers/accepting-collective-responsibility-for-the-future/

Gardiner, S.M., & Fragnière, A. (2018). The Tollgate Principles for the governance of geoengineering: Moving beyond the Oxford principles to an ethically more robust approach. *Ethics, Policy & Environment* 21(2): 143–174.

Gardiner, S.M. (2019a). Motivating (or baby-stepping towards) a Global Constitutional Convention for Future Generations. *Environmental Ethics* 41: 199–220.

Gardiner, S.M. (2019b). The environment and geoengineering. In: D. Edmonds (Ed.), *Ethics and the Contemporary World.* New York: Routledge.

Gardiner, S.M. (2019c). Ethics and geoengineering: An overview. In L. Valera & J.C.C. Zenobi (Eds.), *Global Changes: Ethics, Politics and Environment in the Contemporary Technological World*. New York: Springer.

Gardiner, S.M. (2021a). Intergenerational climate extortion: On making future generations pay for mitigation (and maybe everything else?). *Rivista di Filosofia del Diritto (Review of Legal Philosophy)* 2(2021): 269–284

Gardiner, S.M. (2021b). Debating Climate Ethics revisited. *Ethics, Policy & Environment* 24(2): 89–111.

Gardiner, S.M. (In Press). Climate change and the intergenerational arms race. In B. Williston (Ed.) *Environmental Ethics for Canadians*. Oxford: Oxford University Press.

Gardiner, S.M. (In Preparation). Individual climate responsibility and choosing lives.

Gardiner, S.M., & J. Lawson. (2021). Falling on your own feasibility sword? Challenges for climate policy based on "simple self-interest." In S. Kenehan & C. Katz (Eds.), *Global Climate Change and Feasibility*. New York: Rowman & Littlefield.

Gardiner, S.M., & Weisbach, D.A. (2016). *Debating Climate Ethics*. Oxford: Oxford University Press.

Gilbert, D. (2006, July 2). If only gay sex caused global warming. *Los Angeles Times*. Available at: www.latimes.com/archives/la-xpm-2006-jul-02-op-gilbert2-story.html

Glennie, J., Gulrajani, N., Sumner, A., & Wickstead, M. (2020). A proposal for a new universal development commitment. *Global Policy* 11(4): 478–485.

Goodin, R.E. (2010). Selling environmental indulgences. In S. Gardiner, S. Caney, D. Jamieson, and H. Shue (Eds.), *Climate Ethics: Essential Readings*. Oxford: Oxford University Press.

Goodman, S.G. (2019, April 15). Inequality fuels rage of 'Yellow Vests' in equality-obsessed France. *The New York Times*. Available at: www.nytimes.com/2019/04/15/business/yellow-vests-movement-inequality.html

Green, M.J. (1974, May 26). Deciding on utilities: Public or private? *The New York Times*. Available at: www.nytimes.com/1974/05/26/archives/deciding-on-utilities-public-or-private-con-ed-has-taken-a-step.html

Griffiths, J. (2020, December 3). China to expand weather modification program to cover area larger than India. *CNN World*.

Grohol, J.M. (2016, May 17). 15 common defense mechanisms. *PsychCentral*. Available at: https://psychcentral.com/lib/15-common-defense-mechanisms (Accessed: October 6, 2021).

Grubler, A., Wilson, C., Bento, N., Boza-Kiss, B., Krey, V., McCollum, D.L., Rao, N.D., Riahi, K., Rogelj, J., Stercke, S.D., Cullen, J., Frank, S., Fricko, O., Guo, F., Gidden, M., Havlk, P., Huppmann, D., Kiesewetter, G., Rafaj, P., Schoepp, W., & Valin, H. (2018). A low energy demand scenario for meeting the sustainable development goals without negative emissions technologies. *Nature Energy* 3: 515–527.

Guha, R. (1989). Radical American environmentalism and wilderness preservation: A Third World critique. *Environmental Ethics* 11(1): 71–83.

Hale, B. (2009). What's so moral about the moral hazard? *Public Affairs Quarterly* 23(1): 1–25.

Hale, B. (2011). Nonrenewable resources and the inevitability of outcomes. *The Monist* 94(3): 369–390.

Hall, S. (2017, July 6). What to believe in Antarctica's great ice debate. *Scientific American*. Available at: www.scientificamerican.com/article/what-to-believe-in-antarctica-rsquo-s-great-ice-debate/

Hamilton, C. (2013). *Earthmasters: The Dawn of the Age of Climate Engineering*. New Haven, CT: Yale University Press.

Hands Off Mother Earth! (2018). *Manifesto Against Geoengineering*. Available at: www.geoengineeringmonitor.org/wp-content/uploads/2018/10/home-new-EN-feb6.pdf (Accessed: 6 October, 2021).

Hansen, J., et al. (2005, June 3). Earth's energy imbalance: Confirmation and implications. *Science* 308(3): 1431–1435.

Hardin, G. (1968, December 13). The tragedy of the commons. *Science* 162(3859): 1243–1248.

Hardin, G. (1974, September). Lifeboat ethics: The case against helping the poor. *Psychology Today*. Available at: www.garretthardinsociety.org/articles/art_lifeboat_ethics_case_against_helping_poor.html

Hari, J. (2009, December 19). The truths Copenhagen ignored. *Independent*. Available at: www.independent.co.uk/voices/commentators/johann-hari/johann-hari-the-truths-copenhagen-ignored-1845114.html

Hartzell, L. (2011). Responsibility for emissions: A commentary on John Nolt's "how harmful are the average American's greenhouse gas emissions?" *Ethics, Policy & Environment* 14: 15–17.

Hartzell-Nichols, L. (2014). Adaptation as precaution. *Environmental Values* 23: 149–164.

Hayhoe, K. (2018). The most important thing you can do to fight climate change: Talk about it. *TED*. Available at: www.ted.com/talks/katharine_hayhoe_the_most_important_thing_you_can_do_to_fight_climate_change_talk_about_it/transcript?language=en (Accessed: October 8, 2021).

Hayhoe, K. (2021). Dr. Katharine Hayhoe Teaches Us How to Talk to People Who Don't Believe in Climate Change. *Jimmy Kimmel Live*. Available at: www.youtube.com/watch?v=LVjmGVufADk (Accessed: October 8, 2021).

Heede, R. (2019). *Carbon majors: Accounting for carbon and methane emissions 1854–2010 methods & results report*. LAP LAMBERT Academic Publishing.

Hempel, C. (1960). Science and human values. In R.E. Spiller (Ed.), *Social Control in a Free Society*. Philadelphia, PA: University of Pennsylvania Press.

Heyward, C. (2013). Situating and abandoning geoengineering: A typology of five responses to dangerous climate change. *Political Science and Politics* 46(1): 23–27.

Heyward, C. (2014). Climate change as cultural injustice. In T. Brooks (Ed.), *New Waves in Global Justice*. New York: Palgrave Macmillan.

Hickman, L. (2016, April 13). Timeline: How BECCS became climate change's "saviour" technology. *Carbon Brief*. Available at: www.carbonbrief.org/beccs-the-story-of-climate-changes-saviour-technology

Hill Jr, T. (1983). Ideals of human excellence and preserving natural environments. *Environmental Ethics* 5(3). Available at: https://rintintin.colorado.edu/~vancecd/phil308/Hill.pdf

Hiller, A. (2011). Morally significant effects of ordinary individual actions. *Ethics, Policy, and Environment* 14(1): 19–21.

Hobbs, R.J., Hallett, L.M., Ehrlich, P.R., and Mooney, H.A. (2011). Intervention Ecology: Applying ecological science in the twenty-first century. *BioScience* 61(6): 442–450.

Hooker, B. (2015). Rule consequentialism. *Stanford Encyclopedia of Philosophy*. Available at: https://plato.stanford.edu/entries/consequentialism-rule/

Horton, J., & Keith, D. (2016). Solar geoengineering and obligations to the global poor. In C.J. Preston (Ed.). *Climate Justice and Geoengineering: Ethics and Politics in the Atmospheric Anthropocene* (pp. 79–92). New York: Rowman & Littlefield.

Hourdequin, M. (2010). Climate, collective action and individual ethical obligations. *Environmental Values* 19: 443–464.

Hourdequin, M. (2019). Climate change, climate engineering, and the 'global poor': What does justice require? *Ethics, Policy & Environment* 21(3): 270–288. Also in S.M. Gardiner, C. McKinnon, & A. Fragnière (Eds.), *The Ethics of "Geoengineering" the Global Climate: Justice, Legitimacy and Governance*. New York: Routledge.

Huddleston, A. (2019, July 17). Happy 200th birthday to Eunice Foote, hidden climate science pioneer. *Climate.gov*. Available at: www.climate.gov/news-features/features/happy-200th-birthday-eunice-foote-hidden-climate-science-pioneer

Huemer, M. (2019). *Dialogues on Ethical Vegetarianism*. New York: Routledge.

Hulme, M. (2014). *Can Science Fix Climate Change? A Case Against Climate Engineering*. Cambridge, UK: Polity.

Hursthouse, R., & Pettigrove, G. (2016). Virtue ethics. *Stanford Encyclopedia of Philosophy*. Available at: https://plato.stanford.edu/entries/ethics-virtue/ (Accessed: October 6, 2021).

Huynen, M., Martens, P., & and Akin, S. (2013). Climate change: An amplifier of existing health risks in developing countries. *Environment Development and Sustainability* 15(6): 1425–1442.

Hyams K., & Fawcett T. (2013). The ethics of carbon offsetting. *WIREs: Climate Change* 4: 91–98.

Inoue, C.Y.A. (2018). Worlding the study of global environmental politics in the Anthropocene: Indigenous voices from the Amazon. *Global Environmental Politics* 18(4): 25–42.

IPCC. (2007). *Climate Change 2007: Synthesis Report. Geneva; IPCC. IPCC. (2014). Synthesis report. Contribution of Working Groups I, II and II to the Fifth Assessment Report of the Intergovernmental Panel on Climate Change.* Geneva: IPCC.

IPCC. (2018). Summary for policymakers. In *Global Warming of 1.5°C*. Available at: www.ipcc.ch/site/assets/uploads/sites/2/2019/05/SR15_SPM_version_report_LR.pdf

IPCC. (2021a). *Climate Change 2021: The Physical Science Basis. Working Group I contribution to the Sixth Assessment Report of the Intergovernmental Panel on Climate Change.* Cambridge, UK: Cambridge University Press.

IPCC. (2021b). Summary for Policymakers. In *Climate Change 2021: The Physical Science Basis*. Cambridge, UK: Cambridge University Press.

Jackson, T. (2011). *Prosperity without Growth: Economics for a Finite Planet*. London: Earthscan.

Jamieson, D. (1992). Ethics, public policy, and global warming. *Science, Technology, & Human Values* 17(2): 139–153.

Jamieson, D. (1996). Ethics and intentional climate change. *Climatic Change* 33(3): 323–336.

Jamieson, D. (2001). Climate change and global environmental justice. In P.E. Edwards and C. Miller (Eds.), *Changing the Atmosphere: Expert Knowledge and Global Environmental Governance*. Cambridge, MA: MIT Press.

Jamieson, D. (2007). When utilitarians should be virtue theorists. *Utilitas* 19(2): 160–183.

Jamieson, D. (2010). Climate Change, responsibility, and justice. *Science and Engineering Ethics* 16: 431–445.

Jamieson, D. (2011). Energy, ethics, and the transformation of nature. In D. Arnold (Ed.) *The Ethics of Global Climate Change* (pp. 16–37). Cambridge, UK: Cambridge University Press.

Jamieson, D. (2013a). Jack, Jill, and Jane in a perfect moral storm. *Philosophy and Public Issues* 3(1): 37–53.

Jamieson, D. (2013b). Some whats, whys and worries of geoengineering. *Climatic Change* 121: 527–537.

Jamieson, D. (2014). *Reason in a Dark Time: Why the Struggle Against Climate Change Failed – And What It Means for Our Future*. Oxford: Oxford University Press.

Johnson, B. (2003). Ethical obligations in a tragedy of the commons. *Environmental Values* 12: 271–287.

Johnson, R., & Cureton, A. (2016). Kant's moral philosophy. *Stanford Encyclopedia of Philosophy*. Available at: https://plato.stanford.edu/entries/kant-moral/ (Accessed: October 6, 2021).

Junger, S. (1997). *The Perfect Storm: A True Story of Men Against the Sea*. New York: W.W. Norton.

Kahan, D.M., & Carpenter, K. (2017). Reply to "Culture versus cognitions is a false dilemma." *Nature Climate Change* 7(7): 457–459.

Kahn Jr., P.H. (2002). Children's affiliations with nature: Structure, development, and the problem of environmental generational amnesia. In P.H. Kahn Jr. & S.R. Kellert (Eds.), *Children and Nature: Psychological, Sociocultural, and Evolutionary Investigations* (pp. 93–116). Cambridge, MA: MIT Press.

Kaposy, C. (2018). *Choosing Down Syndrome: Ethics and New Prenatal Testing Technologies*. Cambridge, MA: MIT Press.

Kavka, G. (1981). The paradox of future individuals. *Philosophy & Public Affairs* 11: 93–112.

Keith, D.W. (2013). *A Case for Climate Engineering*. Cambridge, MA: Massachusetts Institute of Technology Press.

Keith, D.W., Weisenstein, D.K., Dykema, J.A., & Keutsch, F.N. (2016). Stratospheric solar geoengineering without ozone loss. *Proceedings of the National Academy of Sciences of the United States of America* 113(15): 14910–14914.

Kelleher, J.P. (2017). Descriptive versus prescriptive discounting in climate change policy analysis. *Georgetown Journal of Law and Public Policy* 15: 957–977.

Khokhar, T. (2015, November 19). Is the term "developing world" outdated? *World Economic Forum*. Available at: www.weforum.org/agenda/2015/11/is-the-term-developing-world-outdated/.

Kim, I.S., & Milner, H.V. (2021). Multinational corporations and their influence through lobbying on foreign policy. In C.F. Foley (Ed.) *Global Goliaths:*

Multinational Corporations in the 21st Century Economy (pp. 497–536). Washington, DC: The Brookings Institution.

King Jr., M.L. (1963). Letter from a Birmingham Jail. In Gates, H.L., & McKay, N.Y. (Eds.), *The Norton Anthology of African American Literature.* New York: W.W. Norton.

Kingston, E., & Sinnott-Armstrong, W. (2018). What's wrong with joyguzzling? *Ethical Theory and Moral Practice* 21: 169–186.

Klare, M.T. (2020). *All Hell Breaking Loose: The Pentagon's Perspective on Climate Change.* New York: Picador.

Kolbert, E. (2014). *The Sixth Extinction: An Unnatural History.* New York: Henry Holt.

Kolbert, E. (2021). *Under a White Sky: The Nature of the Future.* New York: Random House.

Kovaka, K. (2019). Climate change denial and beliefs about science. *Synthese* 198: 2355–2374.

Kraemer, M.U.G., Reiner, R.C., Brady, O.J., *et al.* (2019). Past and future spread of the arbovirus vectors *Aedes aegypti* and *Aedes albopictus. Nature Microbiology* 4: 854–863.

Kumar, R. (2003). Who can be wronged? *Philosophy & Public Affairs* 31(2): 99–118.

Kul, S., & Strauss, B. (2019). New elevation data triple estimates of global vulnerability to sea-level rise and coastal flooding. *Nature Communications* 10(4844): 1–12.

Lackner, K.S., *et al.* (2016). The promise of negative emissions. *Science* 354(6313): 714.

Latham, J., Bower, K., Choularton, T., Coe, H., Connoly, P., Cooper, G., Craft, T., *et al.* (2012). Marine cloud brightening. *Philosophical Transactions of the Royal Society A: Mathematic, Physical and Engineering Sciences* 370(1974): 4217–4262.

Lavelle, M. (2016, May 23). Crocodiles and palm trees in the Arctic? New report suggests yes. *National Geographic.* Available at: www.nationalgeographic. com/science/article/160523-climate-change-study-eight-degrees

Lawford-Smith, H. (2016). Climate matters pro tanto, does it matter all-things-considered? *Midwest Studies in Philosophy* 40(1): 129–42.

Lear, J. (2008). *Radical Hope: Ethics in the Face of Cultural Devastation.* Cambridge, MA: Harvard University Press.

Lee, K. (1999). *The Implications of Deep Science and Deep Technology for Environmental Philosophy.* New York: Lexington Books.

Lenferna, A., Rossotto, R.D., Tan, A., Gardiner, S.M., & Ackerman, T.P. (2017). Relevant climate response tests for stratospheric aerosol injection: A combined ethical and scientific analysis. *Earth's Future* 5(6): 577–591.

Lenferna, A.L. (2018). *Equitably Ending the Fossil Fuel Era: Climate Justice, Capital and the Carbon Budget.* Doctoral dissertation, University of Washington.

Lenton, T.M., Rockström, J., Gaffney, O., Rahmstorf, S., Richardson, K., Steffen, W., & Schellnhuber, H.J. (2019). Climate Tipping Points: Too Risky to Bet Against. *Nature* 575. 592–595.

Lenzi, D. (2021). On the permissibility (or otherwise) of negative emissions. *Ethics, Policy & Environment* 24(2): 123–136.

Lexico. (n.d.) *Anarchism.* Available at: www.lexico.com/en/definition/anarchism (Accessed: October 5, 2021).

Lexico. (n.d.) Propaganda. Available at: www.lexico.com/definition/Propaganda (Accessed: October 6, 2021).

Lichtenberg, J. (2010). Negative duties, positive duties, and the "new harms." *Ethics* 120: 557–578.

Light, A. (2016). Climate diplomacy. In S.M. Gardiner and A. Thompson (Eds.), *The Oxford Handbook of Environmental Ethics*. Oxford: Oxford University Press.

Low Carbon Monitor. (2018). *Pursuing the 1.5°C Limit: Benefits and Opportunities*. Available at: www.thecvf.org/wp-content/uploads/low-carbon-monitor-lowres. pdf (Accessed: October 6, 2021).

Lowell, J. (2012). Managers and moral dissonance: Self justification as a big threat to ethical management? *Journal of Business Ethics* 105(1): 17–25.

Lukacs, M. (2017, July 17). Neoliberalism has conned us into fighting climate change as individuals. *The Guardian*. Available at: www.theguardian. com/environment/true-north/2017/jul/17/neoliberalism-has-conned-us-into-fighting-climate-change-as-individuals

Lynas, M. (2011). *The God Species: Saving the Planet in the Age of Humans*. National Geographic Society.

Lynas, M. (2020). *Six Degrees: Our Future on a Hotter Planet*. New York: HarperCollins.

Maclean, D. (2019). Climate complicity and individual accountability. *The Monist* 102(1): 1–21.

Maltais, A. (2013). Radically non-ideal climate politics and the obligation to at least vote green. *Environmental Values* 22(5): 589–608.

Mann, M.E., & Kump, L.R. (2015). *Dire Predictions: Understanding Climate Change*. New York: Penguin Random House.

Marks, E., Hickman, C., Pihkala, P., Clayton, S., Lewandowski, R.E., Mayall, E.E., Wray, B., Mellor, C., & van Susteren, L. (2021). Young people's voices on climate anxiety, government betrayal and moral injury: A global phenomenon. *The Lancet*. Available at: https://papers.ssrn.com/sol3/Delivery. cfm/edda27cd-a891-47c3-902c-334ddfdf7f84-MECA.pdf?abstractid=3918 955&mirid=1

Marris, E. (2011). *Rambunctious Garden: Saving Nature in a Post-Wild World*. London: Bloomsbury.

Marris, E. (2020, September 10). The West has never felt so small. *The Atlantic*. Available at: www.theatlantic.com/science/archive/2020/09/west-coast-fires-are-making-me-claustrophobic/616252/

Marshall, M. (2020, May 26). Planting trees doesn't always help with climate change. *BBC Future*.

Matthews, D. (2011, August 23). Everything you need to know about the Fairness Doctrine in one post. *The Washington Post*. Available at: www.washing-tonpost.com/blogs/ezra-klein/post/everything-you-need-to-know-about-the-fairness-doctrine-in-one-post/2011/08/23/gIQAN8CXZJ_blog.html

McKibben, B. (1989). *The End of Nature*. New York: Anchor Books.

McKinnon, C. (2016). Should we tolerate climate change denial? *Midwest Studies in Philosophy* 40(1): 205–216.

McKinnon, C. (2019). Sleepwalking into lock-in? Avoiding wrongs to future people in the governance of solar radiation management research. *Environmental Politics* 38(3): 441–459.

McShane, K. (2017). Values and harm in loss and damage. *Ethics, Policy & Environment* 20(2): 129–142.

McSweeny, R., & Tandon, A. (2020). Global carbon project: Coronavirus causes "record fall" in fossil-fuel emissions in 2020. *Carbon Brief*. Available at: www.carbonbrief.org/global-carbon-project-coronavirus-causes-record-fall-in-fossil-fuel-emissions-in-2020

Mersha, S. (2018). Black lives and climate justice: Courage and power in defending communities and Mother Earth. *Third World Quarterly* 39(7): 1421–1434.

Meyer, L. (2013). Why historical emissions should count. *Chicago Journal of International Law* 13(2): 597–614.

Meyer, L., & Roser, D. (2009). Enough for the future. In A. Gosseries and L. Meyer (Eds.), *Intergenerational Justice* (pp. 219–248). Oxford: Oxford University Press.

Miceli, M., & Castelfranchi, C. (1998). Denial and its reasoning. *British Journal of Medical Psychology* 71: 139–152.

Mills, C. (2001). Black trash. In L. Westra and B. Lawson II. (Eds.). *Faces of Environmental Racism: Confronting Issues of Global Justice* (pp. 73–91). New York: Rowman & Littlefield.

Mintz-Woo, K. (2019). Principled utility discounting under risk. *Moral Philosophy and Politics* 6(1): 89–112.

Mintz-Woo, K. (2021). A philosopher's guide to discounting. In M.B. Budolfson, T. McPherson, & D. Plunkett (Eds.), *Philosophy and Climate Change*. Oxford: Oxford University Press.

Mitz-Woo, K. (In Preparation). Will carbon taxes help address climate change?

Moellendorf, D. (2006). Hope as a political virtue. *Philosophical Papers* 35(3): 413–433.

Moellendorf, D. (2014). *The Moral Challenge of Dangerous Climate Change*. Cambridge: Cambridge University Press.

Morrow, D.R. (2014). Starting a flood to stop a fire? Some moral constraints on solar radiation management. *Ethics, Policy & Environment* 17(2): 123–138.

Morrow, D.R. (2017). Fairness in allocating the global emissions budget. *Environmental Values* 26: 669–691.

Morrow, D.R., Thompson, M.S., Anderson, A., Batres, M., Buck, H.J., Dooley, K., Geden, O., Ghosh, A., Low, S., Njamnshi, A., Noël, J., Táíwò, O.O., Taliti, S., & Wilcox, J. (2020). Principles for thinking about Carbon Dioxide Removal in just climate policy. *One Earth* 3(2): 150–153.

Moses, A. (2020, June 8). "Collapse of civilization is the most likely outcome": Top climate scientists. *Resilience*. Available at: www.resilience.org/stories/2020-06-08/collapse-of-civilisation-is-the-most-likely-outcome-top-climate-scientists/

Mutu, M. (2018). Behind the smoke and mirrors of the Treaty of Waitangi claims settlement process in New Zealand: No prospect for justice and reconciliation for Māori without constitutional transformation. *Journal of Global Ethics* 14(2): 208–221.

National Academy of Science. (2017). *Valuing Climate Damages: Updating Estimation of the Social Cost of Carbon Dioxide*. Washington, DC: National Academies Press.

National Aeronautics and Space Administration. (2020). Land-ocean temperature index. Available at: https://data.giss.nasa.gov/gistemp/graphs/graph_data/

Global_Mean_Estimates_based_on_Land_and_Ocean_Data/graph.txt (Accessed: October 6, 2021).

National Oceanic and Atmospheric Administration. (2021). Trends in atmospheric carbon dioxide. *Global Monitoring Laboratory*. Available at: www.esrl.noaa.gov/gmd/ccgg/trends/mlo.html (Accessed: October 6, 2021).

National Park Service. (n.d.) Yellowstone National Park Protection Act (1872). Available at: www.nps.gov/yell/learn/management/yellowstoneprotection-act1872.htm (Accessed: October 6, 2021).

Nenquimo, N. (2020, October 12). This is my message to the western world – your civilization is killing life on Earth. *The Guardian*. Available at: www.theguardian.com/commentisfree/2020/oct/12/western-worldyour-civilisation-killing-life-on-earth-indigenous-amazon-planet

Nerlich, B., & Jaspal, R. (2012). Metaphors we die by? Geoengineering, metaphors, and the argument from catastrophe. *Metaphor and Symbol* 27(2): 131–147.

New, M., Liverman, D., Schroder, H., & Anderson, K. (2011). Four degrees and beyond: The potential for a global temperature increase of four degrees and its implications. *Philosophical Transactions of The Royal Society* 369. Available at: https://doi.org/10,1098/rsta.2010.0303

Nicholas, N. (2021). *Under The Sky We Make: How to Be Human in a Warming World*. New York: G.P. Putnam's Sons.

NOAA. (2020, May 28). Warming influence of greenhouse gases continues to rise, NOAA finds. *NOAA Research News.* Available at: https://research.noaa.gov/article/ArtMID/587/ArticleID/2626/Warming-influence-of-greenhouse-gases-continues-to-rise-NOAA-finds#:~:text=Research%20Headlines%2C%20Climate-,Warming%20influence%20of%20greenhouse%20gases%20continues%20to%20rise%2C%20NOAA%20finds,analysis%20released%20by%20NOAA%20scientists

Nolt, J. (2011a). Greenhouse gas emission and the domination of posterity. In D. Arnold (Ed.), *The Ethics of Global Climate Change* (pp. 60–76). Cambridge: Cambridge University Press.

Nolt, J. (2011b). How harmful are the average American's greenhouse gas emissions? *Ethics, Policy & Environment* 14(1): 3–10.

Nolt, J. (2013). Replies to critics of 'how harmful are the average American's greenhouse gas emissions? *Ethics, Policy & Environment* 16(1): 11–119.

Nordhaus, W.D. (2007). A review of "the stern review on the economics of climate change." *Journal of Economic Literature* 45(3): 686–702.

Nordhaus, W.D. (2015). Climate clubs: Overcoming free-riding in international climate policy. *American Economic Review* 105(4): 1339–1370.

Nuccitelli, D. (2020, July 30). The Trump EPA is vastly underestimating the cost of carbon dioxide pollution to society, new research finds. *Yale: Climate Connections*.

Nussbaum, M. (2003). Capabilities as fundamental entitlements: Sen and social justice. *Feminist Economics* 9 (2/3): 33–59.

Obst, A.R. (In Preparation, 1). Why you should admit why you shouldn't emit (or contribute to other new harms).

Obst, A.R. (In Preparation, 2). *Wilderness for Wildness: Saving the Wild in a Post-Natural World*.

OECD. (2021). *Recommendation of the Council on Guiding Principles concerning International Economic Aspects of Environmental Policies.* Paris: OECD.

Olopade, D. (2014, February 28). The end of the 'Developing World.' *The New York Times.* Available at: www.nytimes.com/2014/03/01/opinion/sunday/forget-developing-fat-nations-must-go-lean.html

Opperman, R. (2019). A permanent struggle against an omnipresent death: Revisiting environmental racism with Frantz Fanon. *Critical Philosophy of Race* 7(1): 58–79.

Oreskes, N. (2004). The scientific consensus on climate change. *Science* 306(5702): 1686.

Oreskes, N. (2019). *Why Trust Science?* Princeton, NJ: Princeton University Press.

Oreskes, N., & Conway, E.M. (2010). *Merchants of Doubt: How a Handful of Scientists Obscured the Truth on Issues from Tobacco Smoke to Climate Change.* London: Bloomsbury Press.

Ostrom, E. (2009). A general framework for analyzing sustainability of social-ecological systems. *Science* 235(5939): 419–422.

Oswald, Y., Owen, A., & Steinberger, J.K. (2020). Large inequality in international and intranational energy footprints between income groups and across consumption categories. *Nature Energy* 5: 231–239.

Palmer, C. (2019). Assisting wild animals vulnerable to climate change: Why ethical strategies diverge. *Journal of Applied Philosophy* 38(2): 179–195.

Pamplany, A., Gordijn, B., & Brereton, P. (2020). The ethics of geoengineering: A literature review. *Science and Engineering Ethics* 26(1): 3069–3119.

Parfit, D. (1983a). Energy policy and the further future: The identity problem. In D. Maclean & P. Brown (Eds.) *Energy and the Future* (pp. 166–179). Totowa, NJ: Rowman & Littlefield.

Parfit, D. (1983b). Energy policy and the further future: the social discount rate. In D. Maclean & P. Brown (Eds.) *Energy and the Future* (pp. 31–37). Totowa, NJ: Rowman & Littlefield.

Parker, L. (2020, August 7). Microplastics have moved into virtually every crevice on earth. *National Geographic.* Available at: www.nationalgeographic.com/science/article/microplastics-in-virtually-every-crevice-on-earth

Peacock, K. (2021). As much as possible, as soon as possible: Getting negative about emissions. *Ethics, Policy & Environment.* Available at: www.tandfonline.com/doi/epub/10.1080/21550085.2021.1904497?needAccess=true

Pearce, D. (1993). *Economic Values and the Natural World.* London: Earthscan.

Pichler, M., Bhan, M., & Gingrich, S. (2021). The social and ecological costs of reforestation: Territorialization and industrialization of land use accompany forest transitions in Southeast Asia. *Land Use Policy* 101(2021). Available at: www.sciencedirect.com/science/article/pii/S0264837720325187/pdfft?md5=a6d3a213e52a9c77677ce47d0e3f15d8&pid=1-s2.0-S0264837720325187-main.pdf

Piurek, R. (2019, October 1). University marking the legacy of IU legend and Nobel Prize winner Elinor Ostrom. *Indiana University.* Available at: https://news.iu.edu/stories/2019/10/iub/inside/01-nobel-prize-winner-elinor-ostrom-legacy-marked.html (Accessed: October 6, 2021).

Pogge, T.W. (2002). *World Poverty and Human Rights.* Cambridge: Polity Press.

Posner, E.A., & Weisbach, D.A. (2010). *Climate Change Justice*. Princeton, NJ: Princeton University Press.

Posner, E.A., & Weisbach, D.A. (2013). International Paretianism: A defense. *Chicago Journal of International Law* 13(2): 347–358.

Preston, C.J. (2012). Beyond the end of nature: SRM and two tales of artificity for the Anthropocene. *Ethics, Policy & Environment* 15(2): 188–201.

Preston, C.J. (2013). Ethics and geoengineering: Reviewing the moral issues raised by solar radiation management and carbon dioxide removal. *WIREs Climate Change* 4: 23–37.

Preston, C.J. (2018). *The Synthetic Age: Outdesigning Evolution, Resurrecting Species, and Reengineering Our World*. Cambridge: Cambridge University Press.

Preston, C.J., & Carr, W. (2019). Recognitional justice, climate engineering, and the care approach. *Ethics, Policy & Environment* 21: 308–323.

Quarcoo, A. (2020). Global democracy supporters must confront systemic racism. *Carnegie Endowment for International Peace*. Available at: https://carnegieendowment.org/2020/07/15/global-democracy-supporters-must-confront-systemic-racism-pub-82298

Rahman, A.A., Artaxo, P., Astrat, A., & Parker, A. (2018). Developing countries must lead on solar geoengineering research. *Science* 22(556): 22–24.

Raterman, T. (2012). Bearing the weight of the world: On the extent of an individual's environmental responsibility. *Environmental Values* 21: 417–436.

Rayner, S., Heyward, C., Kruger, T., Pidgeon, N., Redgwell, C., & Savulescu, J. (2013). The Oxford principles. *Climactic Change* 121(3): 499–512.

Revkin, A.C., & Zeller Jr., T. (December 9, 2009). U.S. negotiator dismisses reparations for climate. *The New York Times*. Available at: www.nytimes.com/2009/12/10/science/earth/10climate.html

Ricke, K.L., Moreno-Cruz, J.B., & Calderia, K. (2013). Strategic incentives for climate geoengineering coalitions to exclude broad participation. *Environmental Research Letters* 8. Available at: stacks.iop.org/ERL/8/014021

Ritchie, H. (2019). How do CO_2 emissions compare when we adjust for trade? *Our World in Data*. Available at: https://ourworldindata.org/consumption-based-co2

Robeyns, I., & Byskov, M.F. (2020). The capability approach. *Stanford Encyclopedia of Philosophy*. Available at: https://plato.stanford.edu/cgi-bin/encyclopedia/archinfo.cgi?entry=capability-approach (Accessed: October 6, 2021).

Robock, A. (2008). 20 reasons why geoengineering may be a bad idea. *Bulletin of the Atomic Scientists* 64(2): 14–18.

Robock, A., Bunzl, M., Kravitz, B., & Stenchikov, G.L. (2010). A test for geoengineering? *Science* 327: 530–531.

Sagoff, M. (1988). *The Economy of the Earth*. Cambridge, UK: Cambridge University Press.

Sagoff, M. (2014). Climate matters: Ethics in a warming world, by John Broome (Review). *Mind* 123(489): 194–197.

Saks, E.R. (2008). *The Center Cannot Hold: My Journey Through Madness*. New York: Hyperion.

Samayoa, M. (2020, September 15). Oregon's air is so hazardous it's breaking records. *Oregon Public Broadcasting*. Available at: www.opb.org/article/2020/09/15/oregons-air-is-so-hazardous-its-breaking-records/

Sandel, M. (2005). Should we buy the right to pollute? In M.J. Sandel (Ed.), *Public Philosophy: Essays on Morality in Politics*. Cambridge, MA: Harvard University Press.

Sandler, R. (2010). Ethical theory and the problem of inconsequentialism: Why environmental ethicists should be virtue-oriented ethicists. *Journal of Agricultural and Environmental Ethics* 23: 167–183.

Schipper, L.F. (2020). Maladaptation: When adapting to climate change goes very wrong. *One Earth* 3: 409–414.

Schmidtz, D. (2001). A place for cost–benefit analysis. *Philosophical Issues* 11: 148–171.

Schwartz, T. (1978). Obligations to posterity. In R. Sikkora and B. Barry (Eds.), *Obligations to Future Generations*. Philadelphia, PA: Temple University Press.

Schwenkenbecher, A. (2012). Is there an obligation to reduce one's individual carbon footprint? *Critical Review of International Social and Political Philosophy*, 17(2), 168–188.

Seager, T.P., Selinger, E., & Spierre, S. (2011). Determining moral responsibility for CO_2 emissions: A reply to Nolt. *Ethics, Policy & Environment* 14: 39–42.

Seip, N. (2018, November 2). Our military bases are not ready for climate change. *The Hill*. Available at: https://thehill.com/opinion/national-security/414540-our-military-bases-are-not-ready-for-climate-change/

Sen, A. (1980). Equality of what? In S.M. McMurrin (Ed.) *Tanner Lectures on Human values*. Salt Lake City, UT: University of Utah Press.

Sen, A. (1983). *Poverty and Famines: An Essay on Entitlement and Deprivation*. Oxford: Oxford University Press.

Seven Generations International Foundation. 7th generation principle. *7th Generation International Foundation*. Available at: www.7genfoundation. org/7th-generation/ (Accessed: October 6, 2021).

Shepherd, J., Caldeira, K., Cox, P., Haigh, J., Keith, D., Launder, B., Mace, G., Mackerron, G., Pyle, J., Rayner, S., Redgwell, C., Watson, A., Garthwaite, R., Heap, R., Parker, A., Wilsdon, J., Thomas, D.J., Fowler, D., Lawton, J., Mitchell, J., Oppenheimer, M., Owens, S., & Read, D. (2009). *Geoengineering the Climate: Science, Governance and Uncertainty*. The Royal Society. Available at: https://royalsociety.org/topics-policy/ publications/2009/geoengineering-climate/

Shockley, K. (2020). Living well wherever you are: Radical hope and the good life in the Anthropocene. *Journal of Social Philosophy* (IF1.07): 1–17.

Shue, H. (1993). Subsistence emissions and luxury emissions. *Law & Policy* 15: 39–60.

Shue, H. (1996). Environmental change and the varieties of injustice. In F.O. Hampson and J. Reppy (Eds.), *Earthly Goods: Environmental Change and Social Justice* (pp. 9–29). Ithaca, NY: Cornell University Press.

Shue, H. (1999). Global environment and international inequality. *International Affairs* 75: 531–545.

Shue, H. (2014). *Climate Justice: Vulnerability and Protection*. Oxford: Oxford University Press.

Shue, H. (2015). Historical responsibility, harm prohibition, and preservation requirement: Core practical convergence on climate change. *Moral Philosophy and Politics* 2(1): 7–31.

Shue, H. (2017). Climate dreaming: Negative emissions, risk transfer, and irreversibility. *Journal of Human Rights and the Environment* 8(2). Available at: https://papers.ssrn.com/sol3/Delivery.cfm/SSRN_ID2940987_code1655536. pdf?abstractid=2940987&mirid=1

Simonet, G., Karsenty, A., Newton, P., de Perthuis, C., Schaap, B., & Seyller, C. (2015). REDD+ projects in 2014: An overview based on a new database and typology. *Climate Economics Chair*. Available at: www.chaireeconomiedu climat.org/en/publications-en/information-debates/id-32-redd-projects-in-2014-an-overview-based-on-a-new-database-and-typology/

Singer, P. (2011). *Practical Ethics*. Cambridge: Cambridge University Press.

Singer, P. (2016). *One World Now: The Ethics of Globalization*. New Haven, CT: Yale University Press.

Sinnott-Armstrong, W. (2005). It is not my fault: Global warming and individual moral obligations. In W. Sinnott-Armstrong & R. Howarth (Eds.), *Perspectives on Climate Change: Science, Economics, Politics, Ethics* (pp. 221–253). New York: Elsevier.

Smil, V. (2016, September 26). "Too cheap to meter" nuclear power revisited. *Institute of Electrical and Electronics Engineers Spectrum*. Available at: https://spectrum.ieee.org/too-cheap-to-meter-nuclear-power-revisited

Smith, P.T. (2018). Legitimacy and non-domination in solar radiation management. *Ethics, Policy & Environment* 21(3): 341–361.

Smith, W., & Henly, C. (2021). Updated and outdated reservations about research into stratospheric aerosol injection. *Climatic Change* 164(39): 1–15.

Solnit, R. (2004). *Hope in the Dark*. Edinburgh, SCT: Canongate Books.

Specter, M. (2012, May 7). The climate fixers. *The New Yorker*. Available at: www.newyorker.com/magazine/2012/05/14/the-climate-fixers#:~:text= Is%20there%20a%20technological%20solution%20to%20global%20 warming%3F&text=Late%20in%20the%20afternoon%20on,that%20 typically%20precede%20an%20eruption.

Standing Bear, L. (1933). Indian Wisdom. In *Land of the Spotted Eagle*. Lincoln, NE: University of Nebraska Press.

Stanley, J. (2016). *How Propaganda Works*. Princeton, NJ: Princeton University Press.

Steininger, K., Lininger, C., Meyer, M., & Muñoz, P. (2015). Multiple carbon accounting to support just and effective climate policies. *Nature Climate Change* 6(1): 35–41.

Stern, N. (2006). *Stern review on The Economics of Climate Change: Executive summary*. London: HM Treasury.

Stern, N. (2008). The economics of climate change. *The American Economic Review* 98(2): 1–37.

Stocker, M. (1976). The schizophrenia of modern ethical theories. *The Journal of Philosophy* 73(14): 453–466.

Strauss, B. (2014, May 14). What does the U.S. look like after 3 meters of sea level rise? *Scientific American*. Available at: www.scientificamerican.com/ article/what-does-the-u-s-look-like-after-3-meters-of-sea-level-rise/

Styron, W. (1979). *Sophie's Choice*. New York: Random House.

Táíwò, O.O. (2022). *Reconsidering Reparations*. Oxford: Oxford University Press.

Táíwò, O.O., & Buck, H.J. (2019, January 31). Capturing carbon to fight climate change is dividing environmentalists. *The Conversation*. Available at: https://

theconversation.com/capturing-carbon-to-fight-climate-change-is-dividing-environmentalists-110142

Tampa, V. (2021, June 22). Congo's latest killer is the climate crisis. Inaction is unthinkable. *The Guardian.* Available at: www.theguardian.com/commentisfree/2021/jun/22/congos-latest-killer-is-the-climate-crisis-inaction-is-unthinkable

The White House. (2021, January 27). Remarks by President Biden before signing executive actions on tackling climate change, creating jobs, and restoring scientific integrity. Available at: www.whitehouse.gov/briefing-room/speeches-remarks/2021/01/27/remarks-by-president-biden-before-signing-executive-actions-on-tackling-climate-change-creating-jobs-and-restoring-scientific-integrity/ (Accessed: October 8, 2021).

The World Bank. (2020). CO2 emissions (metric tons per capita). Available at: https://data.worldbank.org/indicator/EN.ATM.CO2E.PC/ (Accessed: April 30, 2022).

Thompson, A. (2009). Responsibility for the end of nature. Or, how I learned to stop worrying and love global warming. *Ethics & the Environment* 14(1): 79–99.

Thompson, A. (2010). Radical hope for living well in a warmer world. *Journal of Agricultural and Environmental Ethics* 23: 43–59.

Thorpe, D. (2019, November 22). Here's how you can offset your carbon footprint for just $20 per month. *Forbes.* Available at: www.forbes.com/sites/devinthorpe/2019/11/22/heres-how-you-can-likely-offset-your-carbon-footprint-for-just-20-per-month/

Tokarska, L.B., Gillett, N.P., Weaver, A.J., Arora, V.K., & Eby, M. (2016). The climate response to five trillion tonnes of carbon. *Nature Climate Change* 6: 851–855.

Tollefson, J. (2021, August 25). Can artificially altered clouds save the Great Barrier Reef? *Nature* 596: 476–478.

Toman, M. (2001). *Climate Change Economics and Policy.* Washington, DC: Resource for the Future.

Tuana, N. (2019). Climate apartheid: The forgetting of race in the Anthropocene. *Critical Philosophy of Race.* 7(1): 1–31.

Tutu, D. (2007). We do not need Climate Change Apartheid in adaptation. In *Human Development Report 2007/2008, Fighting Climate Change: Human Solidarity in a Divided World.* New York: Palgrave Macmillan.

Tutu, D. (2019). Climate change is the apartheid of our times. *Financial Times.* Available at: www.ft.com/content/9e4befae-e083-11e9-b8e0-026e07cbe5b4

Uenuma, F. (2019, December 30). 20 years later, the Y2K bug seems like a joke – because those behind the scenes took it seriously. *Time.* Available at: https://time.com/5752129/y2k-bug-history/

UNEP. (2019). *Executive Summary to the Global Thematic Report on "Healthy Environment and Healthy People."* Available at: www.unep.org/resources/report/executive-summary-global-thematic-report-healthy-environment-and-healthy-people

UNFCCC. (2014). *Climate Change 2014: Impacts, Adaptation, and Vulnerability.* Available at: https://unfccc.int/topics/science/workstreams/cooperation-with-the-ipcc/the-fifth-assessment-report-of-the-ipcc

United States Environmental Protection Agency. (2018). Sources of greenhouse gas emissions. Available at: www.epa.gov/ghgemissions/sources-greenhouse-gas-emissions (Accessed October 6, 2021).

University Corporation for Atmospheric Research. (2021). History of climate science. https://scied.ucar.edu/learning-zone/how-climate-works/history-climate-science-research

Urban, M. (2015). Accelerating extinction risk from climate change. *Science* 348(6234): 571–573.

van der Linden, S., Maribach, E., Cook, J., Leiserowitz, A., Ranney, M., Lewandowsky, S., *et al.* (2017). Culture versus cognition is a false dilemma. *Nature Climate Change* 7(7): 457.

Victor, D.G. (2011). *Global Warming Gridlock: Creating More Effective Strategies for Protecting the Planet*. Cambridge, UK: Cambridge University Press.

Vidal, J. (2019, August 2). Offsetting carbon emissions: "It has proved a minefield." *The Guardian*. Available at: www.theguardian.com/travel/2019/aug/02/offsetting-carbon-emissions-how-to-travel-options

Vitousek, P.M., Mooney, H.A., Lubchenco, J., & Melillo, J.M. (1997). Human domination of earth's ecosystems. *Science* 277(5325): 494–499.

Vogel, S. (2015). *Thinking Like a Mall: Environmental Philosophy After the End of Nature*. Cambridge, MA: MIT Press.

Wallimann-Helmer, I., Meyer, L., Mintz-Woo, K., Schinko, T., & Serdeczny, O. (2019). The ethical challenges in the context of climate loss and damage. In R. Mechler, L. Bouwer, T. Schinko., S. Surminski, J. Linnerooth-Bayer (Eds.), *Loss and Damage from Climate Change: Concepts, Methods and Policy Options*. New York: Springer.

Wang, E.W. (2019). *The Collected Schizophrenias*. Minneapolis, MN: Graywolf Press.

Wang, P., Deng, X., Zhou, H., & Yu, S. (2019). Estimates of the social cost of carbon: A review based on meta–analysis. *Journal of Cleaner Production* 209: 1494–1507.

Wapner, P. (2010). *Living Through the End of Nature: The Future of American Environmentalism*. Cambridge, MA: The MIT Press.

Watene, W. (2016). Valuing nature: Māori philosophy and the capability approach. *Oxford Development Studies* 44(3): 287–296.

Watt-Cloutier, S. (2010). The Inuit right to culture based on ice and snow. In K.D. Moore and M.P. Nelson (Eds.), *Moral Ground: Ethical Action for a Planet in Peril*. San Antonio, TX: Trinity University Press.

Weinberg, R. (2008). Identifying and dissolving the non-identity problem. *Philosophical Studies* 137(3): 3–18.

Weisbach, D.A. (2016). The problems with climate ethics. In S.M. Gardiner & D.A. Weisbach (Eds.), *Debating Climate Ethics*. New York: Oxford University Press.

Welz, A. (2009). Emotional scenes at Copenhagen: Lumumba Di-Aping @ Africa Civil Society Meeting. Available at: http://adamwelz.wordpress.com/2009/12/08/emotional-scenes-at-copenhagen-lumumba-di-aping-africa-civil-society-meeting-8-dec-2009/ (Accessed: October 6, 2021)

Wenar, L. (2017). *Blood Oil: Tyrants, Violence, and the Rules that Run the World*. Oxford: Oxford University Press.

Werrell, C.E., & Femia, F. (2015, February 12). Climate change as a threat multiplier: Understanding the broader nature of the risk. *The Center for Climate and Security BRIEFER* 25. Available at: https://climateandsecurity.org/

wp-content/uploads/2012/04/climate-change-as-threat-multiplier_under-standing-the-broader-nature-of-the-risk_briefer-252.pdf

Whyte, K.P. (2011). The recognition dimensions of justice in Indian country. *Environmental Justice* 4(4): 199–205.

Whyte, K.P. (2016). Indigenous peoples, climate change loss and damage, and the responsibility of settler states. *SSRN Electronic Journal*. Available at: https://papers.ssrn.com/sol3/Delivery.cfm/SSRN_ID2770085_code1439012.pdf?abstractid=2770085&mirid=1

Whyte, K.P. (2017). Indigenous climate change studies: Indigenizing futures, decolonizing the Anthropocene. *English Language Notes* 55(1–2): 153–162.

Whyte, K.P. (2018). Indigeneity in geoengineering discourses: Some considerations. *Ethics, Policy & Environment* 21(3): 289–307.

Whyte, K.P. (2019). Way beyond the lifeboat: An indigenous allegory for climate justice. In: K.K. Bhavnani, J. Foran, P.A. Kurian, & D. Munshi (Eds.), *Climate Futures: Re-imagining Global Climate Justice*. London: Zed Books.

Wood, R., & Ackerman, T.P. (2013). Defining success and limits of field experiments to test geoengineering by marine cloud brightening. *Climatic Change* 121: 459–472.

Woodson, A., & Gardiner, S. (2019). Climate change, intergenerational ethics, & political responsibility, with Stephen Gardiner. *Carnegie Council for Ethics in International Affairs*. Available at: www.carnegiecouncil.org/studio/multimedia/20191003-climate-change-intergenerational-ethics-political-responsibility-stephen-gardiner (Accessed: October 6, 2021).

Woodward, J. (1986). The non-identity problem. *Ethics* 96: 804–831.

World Bank Group. (2016). CO_2 emissions (metric tons per capita)- United States. Available at: https://data.worldbank.org/indicator/EN.ATM.CO2E.PC?locations=US&most_recent_year_desc=false (Accessed: October 6, 2021).

Wuebbles, D.J., Fahey, D.W., Hibbard, K.A., *et al.* (2017). Our globally changing climate. In *Climate Science Special Report: Fourth National Climate Assessment, Volume I*. U.S. Global Change Research Program.

WWF. (2018). *Living Planet Report – 2018: Aiming Higher*. M. Grooten and R.E.A. Almond (Eds.). Gland, Switzerland.

X, Malcolm. (1964, May 24). *Remarks at a Militant Labor Forum Symposium*. Available at www.youtube.com/watch?v=Ux_zQnD0WfY

Young, I.M. (2006). Katrina: Too much blame, not enough responsibility. *Dissent Magazine*. Available at: www.dissentmagazine.org/article/katrina-too-much-blame-not-enough-responsibility

Young, I.M. (2011). *Responsibility for Justice*. Oxford: Oxford University Press.

Zwally, H.J., Li, J., Robbins, J.W., Saba, J.L., Yi, D., & Brenner, A.C. (2015). Mass gains of the Antarctic Ice Sheet exceed losses. *Journal of Glaciology* 61(230): 1019–1036.

Index

Made in United States
Troutdale, OR
01/04/2025

27609266R00206